Shellfish for the Celestial Empire

Shellfish for the Celestial Empire

The Rise and Fall of Commercial Abalone Fishing in California

Todd J. Braje

THE UNIVERSITY OF UTAH PRESS
Salt Lake City

The Defiance House Man colophon is a registered trademark of the University of Utah Press. It is based on a four-foot-tall Ancient Puebloan pictograph (late PIII) near Glen Canyon, Utah.

20 19 18 17 16 1 2 3 4 5

Library of Congress Cataloging-in-Publication Data

Name: Braje, Todd J., 1976–
Title: Shellfish for the celestial empire : the rise and fall of commercial abalone fishing in California / Todd J. Braje.
Other titles: Rise and fall of commercial abalone fishing in California
Description: Salt Lake City : The University of Utah Press, [2016] | Includes bibliographical references and index.
Identifiers: LCCN 2016005205| ISBN 9781607814962 (pbk. : alk. paper) | ISBN 9781607814979 (ebook)
Subjects: LCSH: Abalone fisheries—California—History. | Abalones—California. | Fishers—California—History. | Chinese Americans—California—History. | China—Emigration and immigration—History—19th century. | Black abalone—California—Channel Islands—History. | Channel Islands (Calif.)—Antiquities. | Channel Islands (Calif.)—History.
Classification: LCC SH371.52.U6 B | DDC 338.3/7243209794—dc23
LC record available at http://lccn.loc.gov/2016005205

Printed and bound by Edwards Brothers Malloy, Inc., Ann Arbor, Michigan.

Contents

Figures

Tables

Preface and Acknowledgments

I began my archaeological studies of the Northern Channel Islands nearly 15 years ago after packing my car with all my worldly possessions and making the 3,000-mile trek from Gainesville, Florida, to Eugene, Oregon. The four years I spent studying at the University of Oregon, completing a doctoral dissertation, and conducting fieldwork on San Miguel Island were some of the most intellectually stimulating, challenging, and rewarding of my life. I intended to explore the deep Native American history of the islands, guided by a historical ecological framework; I left with interests in much more. I found an island, a place, and an archaeological record filled with stories that I wanted to interpret and tell. I became fascinated with historical archaeology, the history of nineteenth-century Chinese immigrants in California, and issues of epistemology, and I struggled to find their intersections with Native American histories and historical ecology.

I have been fortunate to continue my research and field studies on California's Channel Islands. I expanded to new islands and developed classes at universities in Oregon and California that helped me piece together an integrated narrative. I explored the connections between seemingly unrelated topics—precolumbian Vikings in the New World and Chinese history, modern management of nearshore fisheries and ancient coastal foragers, and historical Chinese abalone fishing and black abalone outplanting. The results are a book that ranges widely, covering topics as diverse and ostensibly disparate as the collapse of global fisheries, the California Gold Rush, the science of deep history, marine ecology, and prehistoric and historical archaeology. My intention is to tell a compelling story about the past and present and the connections therein.

To accomplish these goals, I have had to make some difficult decisions, leaving out details (descriptions, numbers, counts, and other facts) that scientists are trained to include in academic publications and technical reports. Many of my professional colleagues may take umbrage at my level of generalization or colloquial style. My ultimate goal, however, is not to speak directly to my colleagues, although I would be thrilled if they found my book compelling. Archaeologists and other scientists have an obligation to communicate with and disseminate results and interpretations to the public, if for no other reason than our research is often sponsored by public funds. In a world filled with misinformation and junk science, this book is my attempt to demonstrate the relevance of the past to the present, to educate the public on how we go about reconstructing ancient history, and to make it all enjoyable. I have found the task much more difficult, but considerably more exciting, than writing an academic treatise. I hope I have come close to my ambitions.

This project is a culmination of a decade and a half of research on the Northern Channel Islands, and any undertaking of this size can only be accomplished with the help and patience of a vast number of people—too many to mention here. While I am the sole author of this book, I have benefited tremendously from the insights, research, and thought-provoking ideas of many close friends and colleagues.

I must first thank Jon Erlandson. Jon convinced me of the importance of historical abalone processing sites on the Northern Channel Islands during my dissertation research in 2005. Many archaeologists, including myself, were focused on prehistoric Native American sites and gave little regard to the piles of black abalone shells scattered across the island coastlines. The seeds of this book germinated in 2006 while Jon and I were excavating a bulk sample at CA-SMI-558, a nineteenth-century abalone processing site on the south-central coast of San Miguel Island, first recorded by University of California–Santa Barbara archaeologist Mike Glassow. Jon also must be given credit for the first half of the book title. We originally intended to publish an academic journal article with a similar title, but a series of fortunate events left the title unused and available for this book. Jon claims it is his favorite to date—I'm thrilled to be the beneficiary!

I owe a tremendous debt to all the colleagues, scholars, managers, and field assistants who supported my research over the last decade. I am very grateful to Torben Rick, Paul Dayton, Robert DeLong, Julia Costello, Don Morris, Laura Rogers-Bennett, Cynthia Catton, René Vellanoweth, Mike Glassow, and Marla Daily for their support, interest, and (many times) collaboration. Since 2011, Linda Bentz has been especially crucial to my research, bringing an infectious energy to our laboratory research group and an expertise in historical records research.

My field and laboratory research was graciously supported by several funding agencies, including the National Science Foundation (BCS-0613982), the National Park Service, Channel Islands National Park, the Western National Parks Association, the Nature Conservancy, San Diego State University, and the SDSU Area of Excellence Fund. At Channel Islands National Park, Ann Huston, Kelly Minas, Laura Kirn, Ian Williams, Mark Senning, and all the park staff assisted with details large and small, from acquiring archaeological research permits to providing logistical support in the field. At the Nature Conservancy and the UC–Santa Cruz Island Reserve, thanks to Christina Boser, David Dewey, Scott Morrison, and Lyndal Laughrin for their help coordinating research on western Santa Cruz Island.

A number of undergraduate and graduate students and volunteers at the University of Oregon, Humboldt State University, and San Diego State University played a critical role in the field research and laboratory analysis. In particular, Linda Bentz, Breana Campbell, Devin Chouinard, Charles Dickerson, Stephanie Duncan, Hannah Haas, Dustin Kennedy, Ryan Ream, and Joel Wirtz all graciously volunteered hundreds of hours. Their relentless efforts are the muscle behind academic research.

At the University of Utah Press, Reba Rauch has been enormously supportive, patient, and helpful throughout this project. Reba and I laid the groundwork for this book back in 2011. Through missed deadlines, endless drafts, and reimaginations, Reba and the Press have been nothing but encouraging. Thank you!

I am grateful to Mark Warner and an anonymous reviewer for their constructive feedback on an earlier version of this book. Their thoughtful comments were critical for improving the book, and I am very grateful for their time and efforts. Hannah Haas assisted in the revision of several figures. Thanks as well to the staff and managers at the University of Utah Press for their help in the final review and production. A special thanks to Kim Vivier for her careful comments and editing.

Finally, I thank my family support network. At home, my wife, Sopagna, and our two dogs, Sitka and Sigmund, have endured the ups and downs of a life lived by academics. Especially during those early years, it was the steadfast belief that everything would work out that saw us through.

That stubborn determination has allowed us to follow our intellectual passions—whether they be island shell middens or cross-cultural psychology, it has been a wonderful ride so far. Above all, my parents, Craig and Sharon, offered constant encouragement and have always been a source of optimism. Mom and Dad, this one's for you!

CHAPTER 1

Fish and Fisheries in Southern California

A History of Collapse

The hidden histories of people, places, events, and landscapes are often uncovered by researchers rummaging through dusty library shelves, consulting stories and memories passed down through the generations, and scanning old newspapers and primary documents for details buried in unexpected places. Some stories, however, are seemingly lost—no one bothered to write them down, or recognize them as important, or understand their influence in shaping the present. Unrecorded histories are often of the oppressed and the disempowered and had the misfortune of taking place in times and places where their value went unrecognized. These histories have power and meaning and a deep importance for our modern world. The oft-quoted adage is certainly true, as George Santayana famously wrote: "Those who cannot remember the past are condemned to repeat it." But the study of history is more than just about avoiding the mistakes and failures of past peoples. History holds lessons about the present and the future. If we are willing to listen, history can help us better understand the world we live in today and where we might be headed in the not-so-distant future. Such is the case with the history of Chinese immigrants who ushered in the birth of commercial abalone (*Haliotis* spp.) fishing in California. Much of their decades-long story was never written down and has been forgotten or marred by a haze of discrimination and racism.

Historical abalone fishing along California's shore began when the newly minted state was caught up in the frenzy of the Gold Rush and its shores and borders were flooded by fortune seekers, domestic and foreign. It was a small group of enterprising Chinese immigrants who first recognized the untapped resources along California's coastline. Their successful efforts to build a commercial abalone harvest and carve out a livelihood were stolen from them by the turn of the twentieth century, however, and their accomplishments relegated to a footnote in California history.

In the mid-nineteenth century, immigrant Chinese abalone collectors altered the palate of a nation by introducing a rarely consumed delicacy. At the time, Euro-Americans viewed abalone as an inedible oddity, consumed only by Asian immigrants in Chinatowns. The texture of abalone was unappealing, and proper cooking techniques were not widely understood. Rocky intertidal shorelines were littered with gigantic abalone up to seven inches in diameter, growing one atop another. This food, highly prized today, was left untouched by Euro-Americans. How did an entire nation overlook such an obvious economic and culinary resource? And what can we possibly learn about the present and future from a marginalized group of Chinese fishermen who lived 150 years ago? Before we answer these questions, let's begin 100 years later when abalone was at its peak popularity.

The 1950s California Shore

Coastal southern California in the 1950s was a land of surf, sand, and sunshine. There was a growing fascination across the country (and the world) with the quintessential southern California lifestyle, and eager participants flooded the beaches from Santa Barbara to San Diego. Surfing, made popular by the famous Hawaiian Duke Kahanamoku in the 1920s (Figure 1.1), was skyrocketing in popularity, and what is believed to be the first surfboard shop, Velzy Surfboards, opened in Manhattan Beach in 1949. Surf music was born in the mid-1950s in Orange County and rapidly became popular, reaching mainstream audiences by 1966, when the Beach Boys released their critically acclaimed album, *Pet Sounds*, and Billboard Hot 100 number-one hit, "Good Vibrations." Fast-food burger joints were all the rage, many featuring drive-ins and carhop service. Beginning in 1948, two particularly successful California restaurants grew to worldwide notoriety, McDonald's, which was founded in San Bernardino, and In-N-Out Burger, founded in Baldwin Park just east of Los Angeles.

The beach and surf culture that began to take shape in midcentury southern California has grown and accelerated, influencing fashion, art,

FIGURE 1.1. Surfing legend Duke Kahanamoku with his longboard on the beach in Los Angeles, California, ca. 1920. *Inset:* Duke Kahanamoku catching a wave at Waikiki Beach, O'ahu, Hawaii, ca. 1910. (Composite by author; both images open access via http://commons.wikimedia.org)

literature, music, film, and terminology. Two surfboards strapped to the top of a Pontiac Woodie with blond teenagers riding in the front seats silhouetted by the setting sun is the southern California of most people's memory or imagination. Surf culture has been so powerful, far-reaching, and enduring that a single piece of surf jargon helped define the personality of the prototypical rebellious adolescent—Bart Simpson and his catchphrase, "Cowabunga!"

Long before surfer culture and Hollywood, the other side of coastal southern California's personality was fishing. By the 1950s, fishing was a way of life in California for not only commercial interests but also sport and recreational anglers. Many coastal Californians spent their weekends, holidays, and times in between with bamboo poles dangled over Santa Monica, Mission Bay, or other coastal piers (Ries 1997), catching rockfish (*Sebastes* spp.) and jacksmelt (*Antherinopsis californiensis*). Even beginners could rent a cheap pole for about fifty cents, dangle the line over the pier and into the ocean, and make the bait dance, with pole pivoted on the railing, using an up-and-down action. It wasn't difficult to come home with something, even if the fish were small.

During the Great Depression, fishing provided penniless families a cheap and quick way to supplement their meals with protein (Ries 1997). Fish of every size and variety were caught and sold along coastal piers during those difficult years. Desperate families could peddle fish for a few extra pennies, and hungry laborers could buy their products for cheap. This cottage industry kept many families alive and established fish as a staple food for generations of Californians.

The historical development of many southern California fisheries followed similar trajectories. Commercial and wide-scale sport harvest began in the mid- to late nineteenth century. Fishing pressure and intensity rose through the "golden years" of the early to mid-twentieth century, until signs of stress and overfishing of many species began to appear in the 1950s to 1970s (e.g., Anderson and Stolting 1953; Love et al. 2002; Ries 1997). Two fisheries provide excellent examples: California rockfish and tuna.

Rockfish are a diverse group of finfish, consisting of about 102 species worldwide, 96 of which are found in the coastal waters of the North Pacific and Gulf of California (Figure 1.2; Love et al. 2002). Rockfish can be recognized by the sharp spines on their dorsal fins. They generally occupy midtrophic levels, preying on smaller fish, small crustaceans, and plankton. Members of many species cluster in shallow waters when they are young, feeding mostly on plankton, and then migrate into deeper waters as they age, where they prey on herring, small rockfish, and crustaceans.

The greatest density and diversity of rockfish are found along the southern California Bight (the curving coastline from Point Conception to San Diego) and waters surrounding the Northern Channel Islands, where 56 species live today (Love et al. 2002). This diversity is likely a result of the region's complex mix of oceanographic currents, countercurrents, and eddies, whose interaction results in water-temperature gradients that form the southern and northern limits for a number of species. Channel Island rockfishes occupy a variety of ecological niches and can be found at water depths ranging from the intertidal to offshore habitats over 200 m deep. The greatest diversity of species along the southern California Bight is found along shallow (30–100 m) and deep-shelf (100–200 m) habitats.

FIGURE 1.2. Canary rockfish (*top*) and China rockfish (*bottom*). (Open access image via http://commons.wikimedia.org)

Unlike most other bony fishes that fertilize external eggs, rockfishes internally fertilize eggs and bear live young. Like most other fish, rockfish are capable of reproducing multiple times during their lives, and some of the longest-lived rockfish may reproduce 100 times or more. There is no evidence that reproductive capacity declines through the rockfish lifetime (Love et al. 2002); instead, older and larger rockfish produce healthier larvae and far more offspring than younger rockfish. This reproductive strategy seems to have evolved as an adaptive response to the variable oceanographic conditions of the northeast Pacific, and the reproductive capacity of older fish may be essential for maintaining healthy populations. The timing of reproductive maturity in rockfish varies by species, with some short-lived species reaching maturity in one to two years and others requiring more than two decades of maturation during life spans that can exceed 100 years.

Although historical rockfishing probably began with the first arrival of Europeans, extensive commercial harvests began in the mid-nineteenth century when several large rocky-outcrop species were sold in San Francisco marketplaces. The market for rockfish, caught mostly by hook-and-line, grew slowly throughout the nineteenth century to moderate commercial importance. At the dawn of the twentieth century, 1.3 million pounds were harvested and sold annually from Washington to California, the vast majority as fresh fish and a minority salted (Love et al. 2002). The fishery remained relatively unchanged until the early 1940s. With low catch prices paid to fishers and the bulk of the West Coast industry centered on halibut (*Paralichthys californicus*), salmon (*Oncorhynchus* spp.), and other more desirable species, rockfish remained a relatively unattractive, impractical, and modestly targeted group of species. In 1943, however, the exploitation of rockfish changed virtually overnight. The balloon trawl, a net enabling more effective harvest from rocky reef habitats, was invented, and rockfish became a central part of the Pacific Coast fishery, although low market prices and limited numbers of fishing vessels continued to keep fishing pressure in check (Braje et al. 2011; Love et al. 2002).

In the early 1960s, the industry began to boom when Russian, and later Japanese, fishers targeted massive populations of rockfish over rocky outcrops using large factory trawlers, tow nets, and roller gear. These new technologies allowed more efficient access to populations along large rock piles that were once inaccessible to intensive net fishing. American fishers were granted U.S. government loans and subsidies to help them compete with their foreign counterparts. California fishers retooled their engines, added vessels to their fleet, and purchased Global Positioning System (GPS) devices. In southern California, Chinese and other Asian immigrants used gillnets to harvest nearshore, rocky-reef species to supply local communities. In addition, as more traditionally popular sport and commercial species such as giant sea bass (*Stereolepis gigas*) and yellowtail (*Seriola lalandi*) became overharvested, the public changed its perception of rockfish, and fishers began to outfit their vessels to exploit this untapped commercial potential. These radical changes to the rockfishery led to dramatic increases in its exploitation along the Pacific Coast in the mid-1970s. The California industry peaked in the early 1980s and remained relatively high into the late 1990s, with the annual value of the fishery exceeding $1 billion (Bloeser 1999; Love et al. 2002).

As has become a familiar story with many of California's (and the world's) most commercially important fisheries, warming ocean temperatures, decades of overfishing, and deterioration of nursery habitats have resulted in stark declines in rockfish numbers and biomass. Overharvesting has been the single greatest culprit in the demise of the rockfishery, and major commercial species, including bocaccio (*Sebastes paucispinis*), canary rockfish (*S. pinniger*), cowcod (*S. levis*), darkblotched rockfish (*S. crameri*), Pacific ocean perch (*S. alutus*), widow (*S. entomelas*), yelloweye (*S. ruberrimus*), and others, are now listed at or below 25 percent of their estimated 1970s biomass (Bloeser 1999). Considering that heavy commercial exploitation began only some 40 years ago, the rate of decline is truly astonishing.

The history of tuna fishing in southern California has followed a similar trajectory (Anderson and Stolting 1953; Ellis 2003). There are 13 different tuna species, six of which are commercially exploited, with members of the genus *Thunnus* being the most popular. In southern California, the most traditionally sought after and heavily exploited have been bluefin (*T. thynnus*), yellowtail (*T. albacares*), and albacore (*T. alalunga*).

Tuna are truly remarkable creatures, all having similar streamlined bodies, pointed heads, tapered tails, and lunate-shaped fins. These traits, combined with their warm-bloodedness (most fish are cold-blooded), allow for incredible speed, endurance, and geographic range. Schools of tuna will range over thousands of miles to exploit baitfish and fuel their unique physiology and continually swimming lifestyle (Ellis 2003). Bluefin tunas make the longest migrations of any fish in the world, and provide sport anglers with exhausting battles of tug-and-pull that can last for hours. Trophy fishers have long prized tunas and compete to mount the biggest on their office walls, no small task considering that the largest can reach lengths of 12 feet and weigh in excess of three-quarters of a ton (Figure 1.3).

The commercial fishery is no less enthusiastic. Tuna is exceptionally popular in sushi restaurants around the world, a trend that is especially evident in Japan. At the Tokyo Metropolitan Central Wholesale Market

FIGURE 1.3. Massive bluefin tuna caught off Santa Catalina Island, California, ca. 1910. *Inset:* A boy stands between two bluefin tunas caught off Santa Catalina Island, ca. 1910. (Composite by author; both images open access via http://commons.wikimedia.org)

(also called the Tsukiji Market), the world's largest quantity and variety of wholesale fish and seafood are sold six days a week (Figure 1.4). Tuna is a major part of the daily auction, and high-end sushi restaurants compete for the headless, tailless, ice-lined prizes, some of extraordinary size and value:

FIGURE 1.4. Some of the daily offerings at Tokyo's Tsukiji Market, including a wide variety of (*A*) fish and shellfish, (*B*) abalone, and (*C*) fish heads. (Photos by T. Rick, October 2010)

> Tokyo, January 5, 2011. A giant bluefin tuna fetched a record 32.49 million yen, or nearly $396,000, in Tokyo on Wednesday, in the first auction of the year at the world's largest wholesale fish market.... The price for the 754-pound tuna beat the previous record set in 2001 when a 445-pound fish sold for 20.2 million yen, a spokesman for Tsukiji market said.... The giant tuna, caught off the coast of northern Japan, was among 538 shipped in from around the world for Wednesday's auction. The record-setting price translates to a whopping 95,000 yen per kilogram, or about $526 per pound. (Associated Press 2011)

The worldwide demand for sushi and sashimi, tuna's popularity as a sport fish, and a variety of other factors have resulted in overfishing of many tuna species, generating major concern among conservationists. The evolution of the southern California tuna fishery and the processes that led to where we are now, however, have a deep and complicated history.

The history of tuna fishing in California began before the arrival of the Spanish, when Native American groups such as the Santa Barbara Chumash fished for yellowfin and other tunas beginning about 3,000 years ago. From boats, likely using seagrass line and abalone shell or composite bone fishhooks, the Chumash captured tunas, billfish, and other deep-water, pelagic species. Since they are large and powerful fish and their successful landing was likely predicated on considerable luck and skill, tunas were not a common part of the Chumash diet. The inherent danger involved in the capture of tunas from indigenous boats (built from either tule reeds or split redwood planks) would have made them a celebrated meal and one that elevated the status of the fisher.

After Spanish arrival and Euro-American settlement in coastal California, the historical tuna fishery remained small and sporadic. During this time, tunas were caught mostly as sport fish with hook and line. The entire industry was suddenly changed, however, when in 1903 the normal run of sardines into the San Pedro area, directly south of Los Angeles, failed to materialize. At the time, sardines were the canned fish of choice and had been commercially packed and sold as such since the 1890s. The sardine failure sent shockwaves through the fishery, and cannery communities and locals scrambled to keep their businesses afloat (Crawford 2009).

One especially industrious southern California cannery owner, Albert Halfill, began experimenting with a replacement for sardines, including rock cod, jewfish, and halibut. Halfill finally settled on canned, steamed albacore tuna. The steaming improved the flavor of the fish, and 700 cases were canned and distributed to wholesalers in that first year (Crawford 2009). Initially, people didn't take to the new product, preferring the more familiar canned salmon produced in facilities to the north. But with increased marketing, canned tuna grew in popularity, and by 1915, species of tuna other than albacore were packaged in southern California, including yellowfin and bluefin. Albacore, the so-called chicken of the sea, was canned as "white meat," and skipjack and yellowfin became "light tuna."

By the 1920s, increased domestic and overseas demand for canned tuna and its popularity as a sport fish combined to result in the explosive growth of tuna fishing in California and worldwide. Large fishing vessels equipped with new ice-lined and insulated fish holds stayed at sea for extended fishing expeditions. In southern California, the year-round availability of skipjack and yellowfin allowed massive fishing fleets to harvest in excess of 100 million pounds of tuna annually by the 1930s. Tens of thousands of southern Californians were linked in some way with the tuna industry, as fishers, factory workers, distributors, sport anglers, or consumers—so much so that the San Diego waterfront became known as the "tuna capital of the world" (Crawford 2009).

Over the next several decades, increased demand, new technologies, growing fishing fleets, and improved harvesting techniques combined to stress West Coast and global tuna populations. Until the 1950s, tuna were captured by hook and line, but increased demand spurred the switch to massive ocean nets that could corral entire schools, wiping out not only the intended targets but also bycatch such as dolphins and sharks that swam with the tuna. By the 1960s, global catches were plummeting—the result of widespread overfishing—and by the early 1980s, all of San Diego's tuna canneries had closed (Crawford 2009; Ellis 2003:28–39).

Unfortunately, the closure of many canneries in California, and around the world, has not stemmed the tide of overfishing. As global tuna populations have plummeted, dock and restaurant prices have risen: a single portion of red, fatty tuna meat, sold as sashimi, can fetch prices upward of $75. Today the tuna fishery rolls on, largely unregulated or at least very lightly regulated, and fish are flown overnight from every corner of the ocean, from Canada to Mexico to New Zealand, to stock the daily auctions at the Tsukiji Market (Figure 1.5). Ocean-based tuna farms are now becoming increasingly popular, despite concerns about genetic manipulation, the potential for unintended environmental consequences, and human health issues associated with eating farmed fish (e.g., Tidwell and Allan 2001; Vita and Marin 2007). Only time will tell if fish farms will cause more harm than good.

Rockfish and tuna represent two of many failed or degraded fisheries in southern California. After over 100 years of evolution in commercial and sport fisheries, California nurtured an identity and economy based, at least in part, on the sea. Overfishing has dealt a heavy blow, and in the last

FIGURE 1.5. A small sample of the number and size of tuna sold daily in the Tsukiji Market. (Open access image via http://commons.wikimedia.org) *Inset:* Processed tuna midsections waiting to be sold at the Tsukiji Market. (Photo by T. Rick, October 2010)

several decades the state has turned to technology, entertainment, and tourism, although fishing remains an important industry for a number of coastal communities. An often overlooked piece of this coastal identity is the commercial and sport gathering of shellfish. Shellfisheries were once (and for some species still are) a thriving part of the California economy.

The most important shellfishery in California today is the sea urchin industry (Kalvass and Rogers-Bennett 2001). Ironically, sea urchins (*Strongylocentrotus* spp.) were once regarded as pests. If urchins are not kept in check, through either human harvesting or predation by sea otters and other predators, populations can explode and will feed on the holdfasts of living kelp. Urchin blooms can deforest kelp in rapid fashion, disrupt local food webs, and create areas of significantly reduced prime productivity. On land, it is analogous to clear-cutting a redwood forest—although, unlike redwoods, kelp can recover and regenerate at incredible speeds. Regardless, kelp deforestation can set off an ecological chain reaction (what biologists call a trophic cascade) that can cause local declines in sea mammals, fish, and shellfish.

The California sea urchin fishery is currently estimated to be worth $75 million per year (Kalvass and Rogers-Bennett 2001). The fishery began only in the early 1970s under a federal program to foster the business, to meet the growing demand for urchin roe in Japanese sushi restaurants, and to limit the potential for devastating urchin blooms (Kalvass and Rogers-Bennett 2001). In California, there are some 400 licensed urchin divers, who primarily target the large red sea urchin (*S. franciscanus*). Fishers can pick urchins of legal size during a restricted season, but there is no agreement on fishing quotas, and annual harvests have decreased after a peak in the late 1980s.

Other shellfish species (turban snails, California mussels, limpets, etc.) are harvested along the California coast as part of recreational fisheries, without any appreciable commercial harvest. The collection is regulated with bag limits and seasonal closings, but beachcombers continue to harvest a diverse range of shellfish for both personal consumption and bait. The most important, controversial, and tumultuous shellfishery in California, however, has been for abalone. Abalone has a deep history in California and was, at one time, as important and emblematic of the California coast as surfing, sun, sand, and fishing.

Historical Abalone Fishing in California

Abalone live in coastal waters worldwide. They can be found along rocky shores in the Pacific, Atlantic, and Indian Oceans, but the greatest concentrations are off the coasts of Australia, Japan, and western North America (California Fish and Game Commission 2001). In the Western Hemisphere, abalone range from Sitka Alaska to southern Baja California and comprise eight species: black (*Haliotis cracherodii*), flat (*H. walallensis*), green (*H. fulgens*), pink (*H. corrugata*), pinto (*H. kamtschatkana*), threaded (*H. assimilis*), red (*H. rufescens*), and white (*H. sorenseni*). (Threaded abalone are sometimes classified as a subspecies of pinto abalone and given the name *H. kamtschatkana assimilus*; pinto are given the name *H. kamtschatkana kamtschatkana*.) These large marine mollusks have one-piece shells that are flattened and ear-shaped. A row of respiratory pores runs along one side of the shell, and a strong muscular foot underneath allows the abalone to adhere itself to rocky surfaces. Abalone are relatively slow to mature and are found in a range of local habitats, depending on the adaptations of the various species (California Fish and Game Commission

2005). Black abalone, for example, prefer the warmer intertidal waters from central California to northern Baja, while red abalone prefer the cooler subtidal waters from southern Oregon to central Baja.

Abalone feed primarily on drift kelp, including detritus from the large and fast-growing giant and bull kelps (*Macrocystis pyrifera*, *Nereocystis luetkeana*) and various red algae. The distribution and abundance of abalone are governed by numerous variables, including water temperature, food availability, rocky substrates, wave energy, predation, disease, and competition for food and space (California Fish and Game Commission 2001:91). Their primary predators are sea otters (*Enhydra lutris*) and humans, but juveniles are preyed on by a wide variety of sea creatures, including sea stars, octopi, crabs, lobsters, and such predatory fishes as sheephead, cabezon, and bat rays (California Fish and Game Commission 2001:91).

Abalone first gained popularity among Euro-Americans in the early part of the twentieth century. Before this time, the giant sea snail was not seen as a viable food source and was largely disregarded by both restaurant owners and home chefs. This is not to say abalone were completely ignored by Euro-Americans. Even as early as 1881, a *Harper's Bazaar* article, "Abalone Jewelry and Its Source," described a thriving domestic trade in abalone shells:

> In Southern California the gathering of haliotis affords employment to a large number of persons, and a considerable commerce has sprung up. There the mollusks are called "abalones"—a word of doubtful Spanish-Indian derivation.... The trade in abalone shells, indeed, is of twice as much importance, financially, as that in the flesh, since it amounted to nearly $90,000 last year. Some Americans also are engaged in this business, and the finishing off of the shells for market is wholly in their hands.... Many of the shells are sent to Europe, there to be polished, with the help of acids, until they shall be as lustrous outwardly as inside, and then are reshipped to the United States to serve as mantel ornaments, soap-basins, match boxes, card-cases, receptacles for flowers, etc. The same work is done to some extent in San Francisco, and many are there manufactured into gold-mounted ear-rings and brooches, shawl pins, and various toilette articles, particularly ladies' high hair-combs of great elegance and costliness. (Anonymous 1881:38–39)

This focus on abalone shell in lieu of the flesh rapidly changed in 1915 when the Panama-Pacific International Exposition (PPIE) was held in San Francisco (Figure 1.6; Lipsky 2005). Designed to celebrate the successful completion of the Panama Canal in 1914 (a project that finished two years ahead of schedule), the fair opened the country's eyes to the Far West. San Francisco, the cultural and economic hub of California, put on display its diverse offerings in culture, cuisine, and entertainment. It was also a message to the country, and perhaps to fellow San Franciscans, that the City by the Bay had recovered from the devastating magnitude-7.9 earthquake and fire that had leveled three-quarters of its homes, businesses, and public buildings in 1906 (Hansen and Hansen 2013). Among the cutting-edge technological exhibits that included a steam locomotive and transcontinental telephone line, there were over 350 varieties of local seafood on display. Included were cooked examples and recipes, as well as a 1,300-seat dining hall to entice the world with California's coastal delicacies.

In the years leading up to the PPIE, abalone was becoming increasingly popular with a small but growing number of foodies outside ethnic Chinese and Japanese communities. In a short article published in *Overland Monthly* in 1913, C.L. Edholm commented:

> A delicacy from the sea, which Americans on the Pacific Coast are just beginning to appreciate, is the abalone, a mollusk which grows to a size of ten inches or more in diameter, within a beautiful iridescent shell.... A few restaurants on the Pacific Coast have undertaken to serve this shell fish to American patrons, but the method of preparing it is far less complicated, and the results are so much more appetizing that the public may become educated up to placing the abalone on the menu within a few years. An excellent way of serving it is to make it into chowder, just as clams are prepared, while another way of serving it is to slice it very thin, pound until tender, and fry like a steak.... Served in either style, it is a most delicious addition to our bill of fare, besides being as wholesome as any other shell fish. When the American public demands this new food, it will greatly increase the industry on the coast, and take it to a great extent out of the hands of the Japanese and Chinese. (Edholm 1913:383–384)

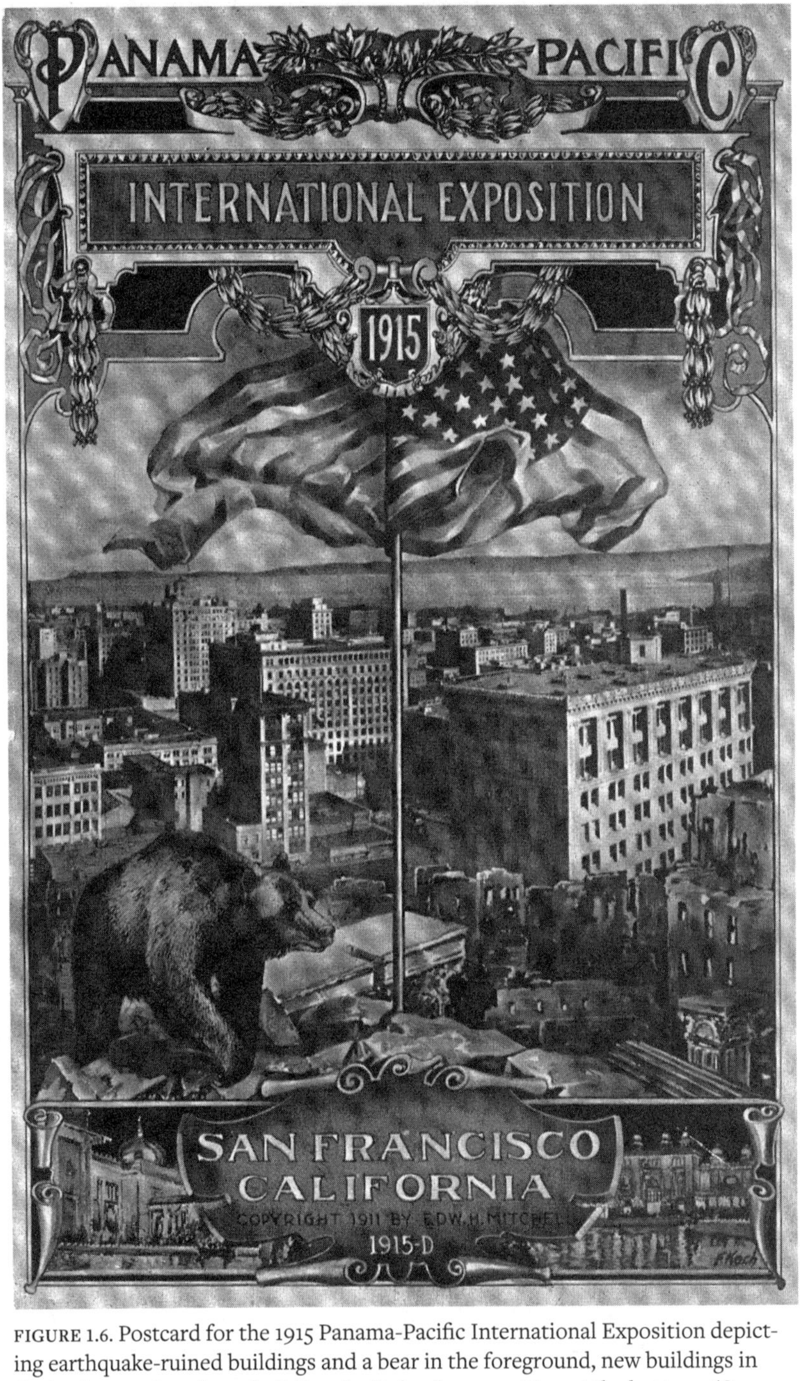

FIGURE 1.6. Postcard for the 1915 Panama-Pacific International Exposition depicting earthquake-ruined buildings and a bear in the foreground, new buildings in the background, and two buildings built for the exposition at the bottom. (Open access image via Library of Congress Prints and Photographs Division, http://www.loc.gov/pictures/resource/ppmsca.19051/)

It didn't seem to matter that the book produced for the PPIE to advertise California's coastal delicacies, Lehner's (1915) *World's Fair Menu and Recipe Book*, made no mention of abalone. After the exhibit, abalone rapidly ascended to culinary prominence, and abalone steaks and chowder became common constituents of restaurant menus.

A fascinating study by the oceanographer Glenn Jones demonstrates just how rapidly abalone rose in popularity. Jones (2008) tallied the available items and associated prices from a wide variety of American menus dating from 1850 to 2006. He found that abalone chowder first appeared on restaurant bills of fare in 1916, and abalone steak in 1923. Earlier appearances are likely, but archival evidence is lacking. Regardless, the point remains: after the PPEI, abalone's popularity skyrocketed. This is not surprising since abalone was a widely available and tasty source of protein. What is surprising, however, is that oysters, clams, mussels, and a variety of other shellfish were already commonly consumed by Euro-Americans, but not abalone. Native Americans had been harvesting red abalone for at least 12,200 years (Erlandson et al. 2011) and black abalone (e.g., Braje 2010:39–52) for nearly as long. In fact, abalone remained a central part of the diets of many California coastal tribes up to and after European arrival.

Euro-Americans certainly recognized abalone's availability and food value. What they seemed to lack was the ability to properly prepare the meat in a way that was satisfactory to the common palate. John Dean Canton pointed this out in an 1891 *Forest and Stream* article: abalone, he wrote, "are not in common use among the Americans, who do not understand the mode of preparing them, while the Mexicans make of them a dish which is admired by all. It is tender, juicy and of fine flavor.... In some parts of the coast they abound in great numbers and several hundred pounds of them are frequently captured at one tide" (Canton 1891:474). The PPEI provided chefs and home cooks the culinary guidance they needed, and by the 1920s, abalone was becoming common fare. Euro-Americans discovered that even a quick preparation could result in a delicious meal, as Canton (1891:474) noted: "To prepare this properly the abalone is removed from the shell, to which it is firmly attached for a space of from 1 to 4 in., according to the size. It is then pounded sufficiently to reduce it almost to a pulp, but not so violently as to break it into pieces. It is then boiled from six to eight minutes, when it is taken from the pot, cut into small pieces, dressed with butter and proper seasoning and served hot. If it is boiled too long it becomes tough and practically not edible." Abalone was added to seafood menus alongside other more traditional items such as grilled swordfish, halibut, and salmon; steamed clams; and sautéed sea bass. By the mid-1920s, abalone chowder and fried abalone steak were common restaurant meals, and the latter cost only about 60 cents (about $6.50 in today's currency; Jones 2008).

Over the next 30 years, abalone only accelerated in culinary popularity. Abalone steaks, burgers, chowder, and a variety of other preparations became increasingly popular in California and throughout the United States. By midcentury, it wasn't uncommon to finish a morning surf session, towel off, and grab a quick (and cheap) abalone burger from the local burger shack along the beach. To meet the growing demand, an entire industry sprouted and blossomed in California, dedicated to the collection and processing of five of the eight species of abalone found in local waters: black, red, pink, green, and white (very little commercial or recreational data is available on flat, pinto, or threaded). Abalone rapidly became one of the mainstays of southern California culture. By the 1950s, California's commercial catch exceeded five million pounds of black, red, and pink abalone. For many Californians, abalone fishing became emblematic of their state and as recognizable as surfing, beaches, tuna, or sardines.

Depletion and Collapse of California Abalone

Despite legislation aimed at protecting and controlling the abalone fishery, declines were evident by the late 1960s, and fishing pressure, both sport and commercial, was taking its toll (Karpov et al. 2000). By the 1970s, commercial landings stabilized at around four million pounds annually, but the number of vessels operating in the commercial fishery had nearly tripled and the market price of abalone steaks rapidly rose, outpacing inflation by an average of 4 percent (Jones 2008:71). It is impossible to know the full extent and impacts of the commercial and sport fishery on California abalone populations because catch records were not consistently recorded by species. Before 1940, only data on red abalone commercial landings were recorded, and records from green, pink, white, and black first appear in 1940, 1944, 1959, and 1956, respectively.

What we do know is that the abalone fishery has seen catastrophic declines since the 1950s. Laura Rogers-Bennett, a marine biologist at

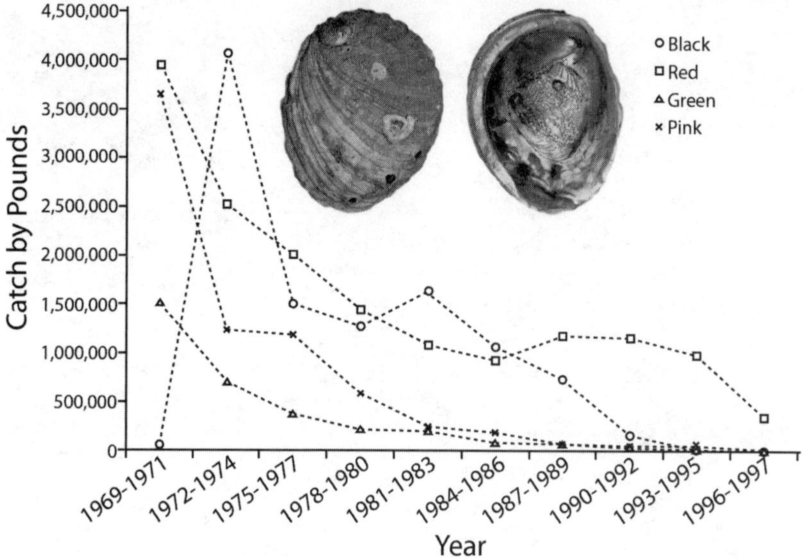

FIGURE 1.7. Commercial and recreational red, black, green, and pink abalone landings in California, 1969–1997. (Based on Rogers-Bennett et al. 2002) *Inset:* A red abalone shell. (Open access image via http://commons.wikimedia.org)

the California Department of Fish and Game, and her colleagues (2002) quantified the historical commercial and sport fishery in California for pink, green, black, white, and threaded abalone. Their results are shocking. After harvest peaks between the 1950s and 1970s, abalone landings fell precipitously (Figure 1.7). During the late 1980s and 1990s, black abalone were hard hit by withering syndrome (WS) disease, which swept through California abalone populations and resulted in a wholesale collapse of the black abalone industry (Haaker et al. 1992). Withering syndrome is a bacterial disease that attacks the digestive track, inhibiting an abalone's ability to produce digestive enzymes. To avoid starvation, the abalone consumes its own body mass, causing its muscular foot to atrophy. The abalone becomes detached from the rock or substrate and typically starves or is eaten by predators.

Although subtidal populations of reds and other abalone were also probably stressed by WS, serial overfishing was likely the primary cause of the collapse (Karpov et al. 2000; Vilchis et al. 2005). By the 1990s, the abalone fishery was in crisis and California began enacting closures. In 1993, all commercial and recreational black abalone fisheries were closed.

FIGURE 1.8. The Abalone Farm, located in Cayucos, California, just outside San Luis Obispo, is the largest abalone aquaculture farm in the United States; red abalone are grown by the thousands here. (Photo by author, December 2008)

Three years later, the commercial and recreational fishery for green, pink, and white abalone was closed (California Fish and Game Commission 2005). In 1997, the next year, the red abalone fishery south of San Francisco Bay was closed.

From a heyday of widespread abundance only a half-century ago, what remains is a small, tightly regulated sport red abalone fishery in northern California. North of San Francisco Bay, divers can collect, three per day and 24 per year, red abalone of legal size (178 mm) during a five-month harvest season without the use of scuba gear. These abalone cannot be legally sold for their meat or shells. Gone is the multimillion-dollar abalone fishery of historical California; today no commercial abalone fishery exists in all of California. Instead, the abalone market is fed exclusively by 13 abalone farms located along coastal central and southern California, where abalone are grown in tanks and fed locally harvested kelp (Figure 1.8).

By any estimation, the wild population of California abalone is only a fraction of its former size (Table 1.1; Rogers-Bennett et al. 2002). Pink, black, and white abalone have been reduced by at least 99 percent of their

TABLE 1.1. Commercial and Recreational Abalone Landings in California during Peak Decades and Estimates of Modern Abundance.

Species	Peak Decade (Commercial)	Commercial Landings	Peak Decade (Recreational)	Recreational Landings	Baseline Estimate	Modern Estimate	% of Peak Abundance
Pink	1950–59	9,318,587	1950–59	N/A	9,300,000	3,200	< 0.1
Green	1966–75	1,473,732	1971–75	30,947	1,500,000	N/A	N/A
Black	1972–81	3,537,126	1972–81	6,729	3,540,000	35,000	< 1.0
White	1969–78	354,973	1971–78	5,503	360,000	< 3,000	< 0.01
Threaded	1971–80	21,066	1971–80	N/A	21,000	< 1,000	N/A

Source: Based on Rogers-Bennett et al. (2002).

historical population peaks (Rogers-Bennett et al. 2002). Black and white abalone are listed as endangered under the U.S. Endangered Species Act, and wild populations are in imminent danger of extinction. The outlook for white abalone is especially dismal. The low densities of wild individuals and repeated reproductive failures make it unlikely that the species will survive and recover without human intervention (Hobday et al. 2001). Too few individuals are left in the wild and they are spread too far apart to effectively reproduce themselves. It will take continued manipulation by the predator that brought them to the brink of extinction (us) to save them from this fate.

Flat, green, pink, pinto, and red abalone are all threatened and being carefully monitored. What this means for the modern abalone consumer is that the same meal of abalone steak from a California coast restaurant that cost an inflation-adjusted $10.56 in the 1950s is today priced at about $75 (Jones 2008). Abalone has disappeared from the dinner plates of the average consumer and can now be found only on those of the privileged.

The Birth of Commercial Abalone Fishing in California

What has been lost in this account of commercial abalone fishing in California are the details of the people who started the industry. In AD 1542 and 1543, Juan Rodríguez Cabrillo, a Portuguese-born explorer, was commissioned by Spain to sail up the California coast. This first European contact with native California led to its eventual colonization by the Spanish over 200 years later in 1769. The Spanish brought with them European crops, cattle, and horses, and established farming and pastoral economies up and down the coast. Spanish towns and missions were erected from San Diego to San Francisco Bay, and local Indians, devastated by the effects of introduced Old World diseases that wiped out somewhere on the order of 90 percent of indigenous Californians, were forced to integrate into European agrarian economies. Coastal tribes were torn from their traditional hunter-gatherer-fisher economies, and many once-important prehistoric fisheries were abandoned. Coastal Indians had relied on the bounty of the sea for millennia and had developed intensive shellfish, fish, and sea mammal hunting economies that were thriving at European contact. For numerous Native American communities, abalone collecting was a central part of traditional lifeways; abalone was used not only for the meat and protein but also for the shells for personal adornments, beads, and fishing tackle such as circular abalone shell fishhooks.

Shortly after occupation by the Spanish, the abalone fishery in California was abandoned. Over the next eight decades, lax predation pressure and human-induced changes to marine food webs allowed abalone populations to flourish. In short order, abalone reached unprecedented sizes and abundances in Pacific Coast waters (Rogers-Bennett et al. 2002; Braje et al. 2009).

It was the Gold Rush nearly 100 years later that triggered a rebirth of abalone fishing in California. Like many other fortune seekers, Chinese immigrants flooded the docks of San Francisco in the mid-nineteenth century, seduced by stories of gold nuggets lining the mountains and streams of the Sierra Nevada. Most Chinese immigrants never realized their riches in the Gold Rush but found instead a land of abundant natural resources and a cauldron of simmering racial tensions. Opposition from hostile American and European gold rushers forced immigrant Chinese workers to the economic fringes of West Coast society. Those who continued to search for gold were relegated to secondhand claims or eked out a livelihood as cooks and laundrymen in roughneck mining camps. Others saw opportunities along California's geographic margins.

In 1853, the first commercial abalone fishing camps were recorded when the *Daily Alta California* described 500 to 600 Chinese immigrants in Monterey gathering intertidal abalone, drying the meat, and shipping it to overseas markets in China. Freed from both human and sea mammal predation for nearly 100 years, abalone stocks were plentiful in coastal California. After the Chinese were pushed out of the gold industry, many realized that there was a fortune to be made in fishing, processing, and exporting the abundant abalone, shrimp, and finfishes of California's coast.

What's more, many immigrant Chinese had come from fishing regions in China. They already had the skills and knowledge base to be successful, along with a niche that needed to be filled. From these seemingly humble beginnings, the Chinese founded the commercial California abalone fishery and were responsible for its growth and expansion over the next several decades. After 30 years of success, however, the U.S. and California governments set about legislating the Chinese out of the abalone industry.

An enormous amount of ink has been spilled on the history, struggles, and successes of Chinese immigrants in North America, no more so than in California, but much less has been written about the early Chinese abalone fishery. This has been the case for three major reasons. First, much of what we know about the history of nineteenth-century American Chinese immigrants has been gleaned from historical documents, occasional diaries, and interviews. Few of these data sources describe, in any great detail, the lifeways and activities of Chinese abalone fishermen. Second, archaeological excavations, the other major source of information, have focused on early Chinatowns, Chinese merchant stores, Chinese laundries, and other more permanently occupied businesses and structures, which are often better preserved and more visible than ephemeral Chinese fishing camps. The fishing camps that have escaped the ravages of coastal erosion, modern development, and time generally are the larger, more permanent communities focused on shrimp, redfish, or other finfishes. Abalone camps were more ephemeral and left smaller, less well preserved archaeological signatures. The slow-growing, largely sedentary nature of abalone required that the fishermen continually move their operations up and down the coast. Finally, the Chinese abalone fishing period was short-lived. Although commercial abalone fishing only grew in popularity and intensity after its initiation by the Chinese, virtually as soon as Euro-Americans recognized its economic potential, they crafted laws to force the Chinese out of the industry (under the auspices of resource conservation).

Fortunately, along with occasional catch records, narratives, and photographs buried deep within historical documents and old newspaper articles, a rather substantial archaeological record of Chinese abalone fishing still exists on southern California's Channel Islands. Scattered along the coastlines of many of the eight offshore islands are the remnants of temporary abalone collecting and drying camps. Timbers, scattered piles of large black abalone shells, and the occasional Chinese pottery, glass bottle, or amorphous metal fragment are all that is left of this once thriving industry. The camps' existence on these relatively isolated islands, away from the highways, strip malls, and population centers of southern California, have left them largely intact. To the untrained eye, the sites may not look like much, but to an archaeologist, they are a treasure trove of information, stories, lifeways, and history of Chinese immigrants. What's more, these sites can tell us about the history and evolution of local intertidal ecosystems, of California abalone, and of kelp forest food webs. They contain lessons about the management and recovery of modern abalone populations and the prospects for a future fishery.

In the following chapters, we'll explore these fascinating records and delve into the history of Chinese abalone fishing. This book is not, however, a tertiary story of nineteenth-century California. It is an analogy for the broader history of Chinese immigrants in America—their struggles, their successes, the institutionalized racism they faced, and the unique ways in which they shaped the identity of our nation. The book also presents a microcosm of our world's fisheries. The story of ecological dysfunction, overharvesting, and eventual collapse is one that can be told for countless species worldwide. The crisis facing Pacific Coast abalone parallels the collapse of many of the most important and productive fisheries around the world. The key to avoiding future crises and restoring our degraded marine ecosystems may be in looking to the past. There are lessons to be learned from history—we just need to know where to look and how to interpret them.

We'll begin our journey not in the nineteenth century but with a detour half a millennium earlier. An increasingly popular myth proposes that a massive fleet of Chinese mariners arrived on the shores of North America's Pacific Coast well before Columbus. This follows a long tradition in popular archaeology of the precolumbian (but post–Native American) arrival of many different groups, ranging from the Lost Tribes of Israel to Druids, Phoenicians, and Atlanteans. In Chapter 2, I explore the good, the bad, and the ugly of precolumbian New World Chinese contacts. I separate the real from the imagined and the scientists from the charlatans. We cannot expect to understand the history of the Chinese in America without first dispelling popular myths about their initial arrival. This chapter also acts an introduction to the science of archaeology. I consider how archaeologists reconstruct the past from a fragmentary archaeological record and explore some of the obstacles facing scientific interpretation and archaeological inquiry.

In Chapters 3 and 4, I explore the history of Chinese immigration to America, the immigrants' mass arrival in San Francisco during the Gold Rush, and the rise of the California abalone fishery. In the next two chapters, I examine the archaeological record of historical abalone fishing on the Channel Islands. Since the archaeological signatures of abalone collecting and processing camps have been largely destroyed on the mainland, island sites offer the best opportunity for direct material study. From this fragmentary record, informed by historical documents, I piece together the lifeways of Chinese abalone collectors, the harsh realities of the industry, and its impact on local abalone populations in southern California. In Chapter 7, I deconstruct the twilight of Chinese abalone fishing, exploring the reasons why the Chinese lost their hold on abalone harvesting and examining California's motivations, after three decades of continual harvest, for legislating the Chinese out of the industry. Was this based on the racist sentiments that had been growing in California and throughout the American West for decades? Or was it to preserve and protect the thriving commercial fishery?

In the final chapter, I track the abalone fishery from the close of the Chinese period to today. I conclude with the lessons of history and how a deep temporal perspective on the evolution of marine ecosystems might help us develop more sustainable marine resource exploitation practices and management. I consider the current state of California abalone populations and the recent debate between fishers and the California Department of Fish and Game over proposals to reopen the red abalone fishery along the Northern Channel Islands. I also explore how tracing black abalone fishing by Native Americans beginning at least 10,000 years ago and commercial fishing by Chinese fishermen 150 years ago can provide a roadmap for effective black abalone restoration efforts. Archaeology, in surprising ways, has something powerful to add, but only if we are willing to listen.

CHAPTER 2

Of Anchors, Admirals, and Artifice

The Chinese before Columbus?

On January 8, 2003, National Public Radio (NPR) aired an interview with Gavin Menzies, a retired Royal Navy submarine commanding officer and best-selling author, on *Morning Edition*, the most popular news radio program in the country. Since its inception in 1979, *Morning Edition* has collected some of broadcasting's highest honors, including the George Foster Peabody Award, which recognizes excellence, achievement, and public service in electronic media, and the Alfred I. duPont–Columbia Award, which honors excellence in broadcast journalism. These are, in my opinion, well-deserved accolades, and *Morning Edition* is often an engaging and informative part of my morning commute.

While *Morning Edition*'s coverage is broad and the program often simply reports on "hot" topics, I think it's safe to assume that inclusion in the radio broadcast provides some legitimacy for the ideas, claims, and "facts" being discussed. The majority of listeners would not think to question the ideas being presented and naturally assume that participation is preceded by intensive fact checking. This is not necessarily a bad thing, as we generally rely on reputable television and radio news programs, books, or newspapers to obtain information on world and current events. While we know that there are well-regarded news outlets (national media companies such as the *New York Times*) and not-so-well-regarded ones (tabloid papers such as the *National Enquirer*—caveat lector!), NPR and *Morning Edition* have reputations for factual and reliable reporting. How, then, do we evaluate what we "learn" from reputable news sources? Do we need to be skeptical of the information we hear on programs such as

Morning Edition? Are there red flags that we can look for and questions we can ask ourselves, even about subjects on which we are not experts? We'll return to these issues later, but for now let's bring our attention back to Gavin Menzies.

In his seven-minute interview, Menzies described the central thesis of his best seller *1421: The Year China Discovered the World.* In a nutshell, Menzies claimed that Chinese mariners discovered, explored, and settled the New World 70 years before Columbus. A lifetime of skills in cartography, geography, and astronavigation, acquired during his many years as a submarine commander, allowed Menzies to discover what countless academics could not. His book is a 650-page treatise that rewrites the history of the world. On *Morning Edition*, he told Bob Edwards: "I came to the astonishing conclusion that a huge Chinese fleet of 800 vessels sailed from China in March 1421 and they had circumnavigated the world and they had created settlements in North and South America, in Australia, New Zealand, across the Pacific, and in the Indian Ocean" (Menzies 2003).

This is astonishing indeed, and fundamentally alters everything we know about the precolumbian history of North and South America, not to mention the history of the seventeenth- and eighteenth-century European discovery and colonization of the Pacific. Edwards asked Menzies if there were direct scientific, archaeological evidence supporting his claims. Menzies responded with a laundry list: "Oh, yeah, there's absolutely stacks of evidence. And there's all sorts of different strands to this evidence, most important in my mind is that the first Europeans who reached the New World met Chinese people. They found Chinese wrecks. There's DNA, there's bodies, there's diseases, which are unique to China in the New World, linguistics, shipwrecks, Chinese porcelain, Chinese jade, Chinese chickens, Chinese rice, all found in the New World by the first Europeans" (Menzies 2003). Given the quantity and quality of this evidence, it's hard (based on the interview) to argue with Menzies, and Edwards did not. Any listener might wonder, "Wow, how did we miss this for so long?" In seven minutes, or 650 pages, everything we once knew about Columbus and the other European explorers was turned on its ear! Should we rewrite the history books? Do we need to take Columbus Day off the list of American holidays? Perhaps so, but not because of Menzies's claims.

How Do We Know What We Know?

Every academically trained archaeologist is well aware that the study of the ancient past can be a breeding ground for wild speculation, gross exaggeration, and pseudoscience. Perhaps it's the legacy of Indiana Jones, tomb raider Lara Croft, or other Hollywood depictions of archaeologists that make the mysteries of the past open for fantastical reconstructions. Since deep history is a relic of the past, and only vague clues to historical events are left behind, there is no step-wise process to unlock the secrets that antiquity holds, right? No. The sobering reality is that archaeologists must play by the rules—the rules of science, that is—in the same way physicists, chemists, and any other scientists do. Archaeologists propose multiple working hypotheses to explain phenomena, set up rigorous experiments or field research to test hypotheses, and reject those that do not hold up to scrutiny. This process of following the scientific method might seem mechanical and simple, but, in fact, science is difficult, time-consuming, and (at least partially) an art form. Carl Sagan (1996:27), one of the foremost scientific thinkers of our time, in his best-selling book about the process and applications of science, *The Demon-Haunted World: Science as a Candle in the Dark*, sums this up nicely: "The scientific way of thinking is at once imaginative and disciplined. This is central to its success. Science invites us to let facts in, even when they don't conform to our preconceptions. It counsels us to carry alternative hypotheses in our heads and see which best fit the facts. It urges on us a delicate balance between no-holds-barred openness to new ideas, however heretical, and the most rigorous skeptical scrutiny of everything—new ideas and established wisdom." Recognizing "mysteries" in the first place and building creative and testable hypotheses takes skill, perseverance, imagination, and ingenuity. Following this process, however, is the only way we can, from one individual to another and across time and space, come to consistent agreement about what we know to be true.

This is not to say that all archaeologists reach the same conclusions when trying to reconstruct the past. Quite the contrary—archaeologists frequently and vigorously disagree about their reconstructions. The annual meeting of professional archaeologists in the United States, the Society for American Archaeology (SAA), is often a lively event where sessions organized around thematic topics sometimes devolve into shouting matches over seemingly mundane archaeological debates (although most disagreements are much more civil and play out over beers or within the pages of academic journals).

If science is objective, how could this be? There are many reasons, which often come down to a simple numbers game: only a small percentage of archaeological sites or past human activities are preserved; only a small percentage of these are recorded in the archaeological record; only a small percentage of these have been scientifically excavated and dated; some regions or time periods are especially poorly known, leaving major gaps in the record; and the archaeological record can be biased by geological or other forces that destroy sites. What archaeologists are left with, then (when we are lucky), amounts to just a few pieces of a complex jigsaw puzzle to reconstruct the incredibly rich picture of past human activities.

What science and the scientific method offer is a logical method for putting the jigsaw puzzle together and a way to evaluate what we think we know. Once we start making claims about the world (past and present) and asking for reasons to accept them, we can expect people to start wondering what reasons are good ones and why we should accept them. From this we get logic, methodology, and epistemology (the study of knowledge). Science, however, can never bring us to a definitive answer. As the archaeologist Kenneth Feder (2008:24) points out in his entertaining text on science and pseudoscience in archaeology, *Frauds, Myths, and Mysteries*:

> In large part, science is a series of techniques used to maximize the probability that what we think we know really reflects the way things are, were, or will be. Science makes no claim to have all the answers or even to be right all the time. On the contrary, during the process of the growth of knowledge and understanding, science is often wrong.... The only claim that we do make in science is that if we honestly, consistently, explicitly, and vigorously pursue knowledge using some basic techniques and principles, the truth will eventually surface and we can truly know things about the nature of the world in which we find ourselves.

Science is not the toil of one particular group of people, or merely a "Western" way of viewing the world. It is the work of people from every culture and every society on earth. It is a way to approach questions about the past, present, and future and is dependent on the accumulated knowledge of every generation of scientists to refine what we know to be true.

Before Columbus, after Native Americans?

Questions about the precolumbian colonization of the New World have long been a source of fascination for professional scholars and amateur history buffs, but unfortunately, some choose to disregard the rules of science. The list of explorers and colonizers purported to have visited North, Central, and South America before Columbus but after Native Americans is too long to enumerate. Table 2.1 offers just a small sample of some of the most famous fantastic and pseudoscientific claims. The authors range from amateur historians and archaeologists to faculty members at academic research institutions. The claims include precolumbian Egyptian, Celt, and Greek mariners; refugees from the lost continent of Atlantis; and extraterrestrial visitors arriving aboard spacecraft. Some of these claims are potentially interesting, and many of the groups *could have* made it to the New World before Columbus, but the question remains: Did they? There is one case of post–Native American, precolumbian contact with the New World that can help us evaluate all of these fantastic claims. It provides a roadmap of how to assess evidence; it demonstrates that anyone can be an effective scientist (if they play by the rules); and it illustrates the importance of hypothesis testing and methodology in archaeology.

Precolumbian Vikings in a New World

For at least 300 years, amateurs and professionals alike have claimed that Vikings landed in the New World well before Columbus. Geographically speaking, there is not much to argue. Greenland is, technically, part of the North American continental landmass, and Vikings first arrived there in AD 985 (more below). The real debate, however, has been over when Vikings first arrived in what is now Canada and the United States.

The list of fraudulent Viking artifacts used as evidence for the precolumbian settlement of North America is long. These were especially abundant in the nineteenth century, when Viking stories were romanticized and commonly used as symbols of ethnic, political, or national identity. Desperate to prove the existence of Vikings on the North American coast, antiquarians scoured the seashore from Labrador to South Carolina for Viking runes and relics. For example, a 40-ton boulder, Dighton Rock,

TABLE 2.1. Popular Pseudoscientific Claims for the Precolumbian Discovery of the New World.

Author	Contact Group	Point of Contact	Time of Contact	Seminal Publication
Churchward, James	Refugees from Mu	New and Old Worlds	12,000 years ago	*The Lost Continent of Mu* (1926)
Deacon, Richard	Prince Madoc and the Welsh	North America	AD 1170	*Madoc and the Discovery of America* (1966)
Donnelly, Ignatius	Atlanteans	New and Old Worlds	9,000 years ago	*Atlantis: The Antediluvian World* (1882)
Fell, Barry	Iberians, Celts, Greeks, Hebrews, Egyptians, among others	North America	3,000–1,500 years ago	*America BC* (1976)
Hancock, Graham	An ancient, lost civilization	New and Old Worlds	10,500 BC	*Fingerprints of the Gods* (1995)
Menzies, Gavin	Chinese	Worldwide	AD 1421–1423	*1421: The Year China Discovered America* (2002)
Van Sertima, Ivan	Africans	Mesoamerica	700 BC and AD 1300	*They Came before Columbus* (1976)
Von Däniken, Erich	Extraterrestrials	Worldwide	Various	*Chariots of the Gods?* (1968)

found in southern Massachusetts in the late seventeenth century and adorned with a series of petroglyphs, was argued in 1837 to contain Viking runes (Figure 2.1; Feder 2008). The search even extended far inland to the American Midwest and Plains. Olof Ohman, a farmer of Swedish descent from Minnesota, claimed to have found a 200-pound stone covered in Viking runes, the Kensington Stone, while preparing a field for plowing in 1898. Similarly, the Heavener Runestone is an inscribed rock located in eastern Oklahoma that has been attributed to a wayward precolumbian Norseman. These, and many others like them, are more likely of Native American construction (Dighton Rock), are of questionable origin, or have been shown to be frauds (the Kensington and Heavener stones).

What we do know about the real Vikings, who they were, their lifeways, and where they settled is remarkable and very different from popular conception. "Viking" is a term likely derived from Old Norse and once

FIGURE 2.1. Dighton Rock, 1893. (Open access image via http://historyproject.ucdavis.edu/ic/) *Inset:* Drawing of the Dighton Rock petroglyphs. (Open access image via http://commons.wikimedia.org)

referred to those who went on expeditions or raids. The Norse, a more appropriate term, refers to the peoples of Scandinavia before Christianization in the late eleventh century. Although the Vikings earned their reputation in Europe as raiders of villages and monasteries along the coastal mainland, the British Isles, and even interior locations along major river drainages, the Norse spent most of their days as farmers, fisherfolk, pastoralists, and expert metalworkers. They also were adept shipwrights and prolific traders with economic ties stretching from what is now Iraq to the Canadian Arctic.

One of the enduring aspects of Viking culture is the Viking saga, an oral tradition of stories of adventure, discovery, life, and death. These tales were passed down through the generations and eventually transcribed hundreds of years after they were first composed. The accuracy of the sagas has been, and continues to be, hotly debated. Some historians have argued that the sagas describe real people and events that can be ground-truthed, while others view the sagas as artistic storytelling.

One long-debated controversy centers on the discovery of a new country that some have claimed is mainland North America. The *Vinland Sagas*, translated by Magnusson and Pálsson (1965), contain two separate but related accounts, the *Saga of the Greenlanders* and the *Saga of Eirik the Red*. According to these sagas, in AD 982 the Althing, an assembly of representatives from all of Iceland's ruling families, met to discuss the case of Erik the Red. Erik was a short-tempered Viking who had arrived in Iceland after being expelled from Norway for murder. In the intervening years, he had married, established a farm, and produced four children—three boys, Leif, Thorvald, and Thorstein, and one daughter, Freydis. Despite his best efforts at a peaceful existence, an argument with a neighbor in Iceland ended in a second (and third) murder. The Althing was tasked with deciding his fate and agreed on a three-year exile.

Erik had heard tales of an undiscovered and unsettled land to the west and set sail. Sailing westward, he happened upon a deserted coastline that led him to a rich landscape dotted with fjords, meadows, and willow and birch forests. In perhaps the greatest real-estate fraud in history, Erik named this new country of mostly snow, ice, and glaciers "Greenland," stating that "people would be much more tempted to go there if it had an attractive name" (Magnusson and Pálsson 1965:77). Three years after his departure, Erik the Red returned to Iceland and recruited a group of colonists to establish the first Norse settlement in Greenland.

During that first year of colonization, a Viking ship sailing from Iceland to Greenland, captained by Bjarni Herjolfsson, was lost in a storm and blown off course. Bjarni and his crew sailed for many days, aimless and hopelessly lost. When the fog lifted, however, they spotted a strange, forested land and at least two other unfamiliar shores. Rather than investigating, Bjarni sailed east, reaching Greenland after four days.

Fascinated by Bjarni's story and the chance to make a name for himself, Leif Erikson, Erik the Red's oldest son, set sail with 35 men due west from Greenland sometime between AD 997 and 1003. According to the *Greenlanders' Saga*, Leif and his crew came first to a land of rock and glaciers (Helluland), sailed to a more temperate wooded bay (Markland), and continued south to winter at a location they named Vinland, likely for the wild grapes they discovered. There they built traditional Viking sod houses and established a base from which the new land could be explored. After the sea ice had retreated, Leif and his crew sailed home to announce the discovery of a land teeming with natural resources.

Leif's brother, Thorvald, was the next to investigate. He explored the new lands to the west for about a year. Unfortunately, he did not fare as well as Leif and died in an altercation with the locals, whom the Vikings named "Skraelings" and described as small and evil-looking. Thorvald was buried on Vinland, and his crew returned to Greenland.

In AD 1022, Thorfinn Karlsefni gathered a group of at least 65 migrants and an exhaustive list of supplies, including livestock, and attempted to establish a permanent Viking colony at Vinland. After spending a year constructing houses, farming, and herding their animals, the colonists abandoned Vinland following a bloody battle with the Skraelings that left two Norsemen and four natives dead. Perhaps stemming from the earlier incident in which Thorvald lost his life, *Eirik the Red's Saga* (Magnusson and Pálsson 1965:100–101) describes the reason for the failed settlement: "Karlsefni and his men had realized by now that although the land was excellent they could never live there in safety or freedom from fear, because of the native inhabitants. So they made ready to leave the place and return home. They sailed off north along the coast."

The *Greenlanders' Saga* and *Eirik the Red's Saga* long offered tantalizing hints of a precolumbian Viking presence in the New World. What was lacking for decades, however, was the material evidence from the archaeological record. Single Norse artifacts in precolumbian archaeological contexts do not constitute conclusive proof, because many of these are found throughout northeastern North America, likely transported there in a variety of ways. Maybe the most famous is the "Norse penny," a perforated coin, perhaps made for use as a pendant or ornament, found within a Native American site in Maine (McKusick and Wahlgren 1980). There are now a number of sites in the Canadian Artic that have produced Norse artifacts, including armor, iron rivets and wedges, smelted copper, and other items that are clearly of European origin. Norse archaeology expert Patricia Sutherland (2000) credits the presence of these to down-the-line trade networks among Native American communities, Indian scavenging of Norse campsites, or direct trade between Norsemen and Native Americans.

Proof of Leif Erikson's and later Viking voyages and colonies in the New World would require finding the settlements themselves as described in the sagas. Two intrepid adventurers rose to the task, the Norwegian explorer Helge Marcus Ingstad and his wife, the archaeologist Anne Stine. After 10 years of background research and investigation, beginning in 1960, Helge went to the field and searched the bays and harbors of

FIGURE 2.2. Reconstructed Viking longhouse at L'Anse aux Meadows. *Inset:* Norwegian explorer Helge Marcus Ingstad, ca. 1942. (Composite by author; both images open access via http://commons.wikimedia.org)

Newfoundland, convinced that this setting was the Vinland of the sagas. He presumed that the sizable colony of Thorfinn Karlsefni would still be preserved since it had been abandoned less than a millennium earlier (and thus was relatively recent by archaeological standards). After talking to local people and consulting old maps, Helge made a remarkable discovery on a marine terrace at Épaves Bay: the remnants of eight typically Norse turf houses (Figure 2.2; Ingstad and Ingstad 2000).

For seven years, Anne Stine directed the excavations at L'Anse aux Meadows, unearthing a remarkable assemblage of Viking artifacts, including a soapstone spindle whorl, iron nails, locally exploited and smelted bog iron, a bronze pin, and many other distinctly Norse artifacts. This was the "smoking broadsword" of precolumbian Viking settlement of the New World. What is perhaps most remarkable, however, is that L'Anse aux Meadows was discovered by an amateur scientist, Helge Ingstad, who effectively used logic, the scientific method, and deduction, without the advantage of any formal archaeological training. Helge, with some help along the way, was able to put to rest one of the enduring mysteries of North American history. There are still debates over whether this location is the settlement of Leif, his brother, or Thorfinn, and we may never know for certain. But what we do know is that the Vikings settled North America some five hundred years before Columbus.

Evaluating Precolumbian Chinese Contact

With this Viking case study in mind, let's turn our attention back to Gavin Menzies and his *1421: The Year China Discovered America*. According to his version of Chinese history, between the years AD 1421 and 1423 four Chinese sailing fleets, made up of 25 to 30 ships carrying at least 7,000 people, visited, mapped, explored, and, occasionally, settled every continent on earth, save Europe. The Chinese captains of each fleet—Hong Boa, Zhou Man, Zhou Wen, and Yang Qing—were trained by the famous eunuch admiral Zheng He and ordered by the third Ming emperor, Zhu Di, to build a maritime empire. According to Menzies (2002:456), the adventure was wildly successful: "These remarkable Chinese admirals rounded the Cape of Good Hope sixty-six years before Dias, passed through the Strait of Magellan ninety-eight years before Magellan, surveyed Australia three centuries before Captain Cook, Antarctica and the Arctic four centuries before the first Europeans, and America seventy years before Columbus."

When the last of the fleet returned to China in AD 1423, the emperor who had dispatched them was dead and his son, Zhu Gaozhi, ordered an end to oceanic exploration. Over the next several hundred years, all the documents detailing the voyages were destroyed and expunged from memory. The Chinese dynastic leaders turned their attention inward, to social and political unrest that swept through the country, and isolated themselves from the outside world.

Despite these difficulties, Menzies includes a catalog of thousands of pieces of supporting evidence and a website providing the latest findings by his team of researchers. The stunning array of data draws from such diverse fields as archaeology, cartography, linguistics, astronomy, geography, history, and genetics. The key piece of evidence that allowed Menzies to track the Chinese voyages and unravel this mystery, however, is a series of fifteenth- and sixteenth-century maps, the most important of which is the Piri Reis map of 1513. Despite the obvious problem that all of these maps *postdate* the Chinese and early European explorations, Menzies contends that they were drawn with knowledge passed from the Chinese to European cartographers and were in the possession of Columbus, Magellan, and other explorers before their departures. Columbus knew where he was destined before he left Spain!

When you start to scratch below the surface of these outlandish claims (and quite a few archaeologists and historians have), Menzies's story quickly unravels. For example, in his *Piri Reis Map of 1513*, the historian Gregory McIntosh (2000) describes the origin and historical context of the Piri Reis map. McIntosh notes that an Ottoman admiral produced the map by combining information from a number of earlier maps (including one by Columbus). Claims that the map includes uncharted locations in North and South America and Antarctica, evidence for Menzies of an earlier Chinese voyage, requires significant imagination and squinting to be convincing. Rather, these locations are nothing more than guesswork by early European cartographers who had an aversion to leaving blank areas on their maps.

Even more disturbing, Menzies ignores scores of surviving documents on the Zheng He expeditions. In a scathing article published in the *Journal of World History*, the University of Arkansas historian Robert Finlay (2004) points out that there were no global Chinese explorations that went undocumented, as claimed by Menzies. Rather, the explorations of Zheng He and his squadrons are well documented; the last of the fleets arrived in China on October 8, 1423, after one year and three months at sea and with no time to undertake even a small portion of Menzies's imagined global explorations. Finlay (2004:242) concludes, "The reasoning of *1421* is inexorably circular, its evidence spurious, its research derisory, its borrowings unacknowledged, its citations slipshod, and its assertions preposterous."

Responding to every piece of Menzies's evidence is outside the scope of this book, although there are efforts to build a large-scale scientific response (see www.1421exposed.com). Rather, let's look at Menzies's three key pieces of evidence for Chinese contacts in coastal North America: (1) a wooden pulley from Nehalem Bay in Oregon; (2) a Chinese junk in Sacramento, California; and (3) submerged Chinese anchors outside Los Angeles. Each of these cases illustrates how Menzies conducts his "research" by ignoring data that do not fit his story and packaging information in clever ways that seem convincing to the casual reader.

A Fifteenth-Century Chinese Wooden Pulley in Oregon?

Nehalem Bay is located along the northern Oregon coast between Manzanita and Nehalem. A majestic land- and seascape, Nehalem Bay State Park is positioned along the narrow, four-mile-long sand spit that separates the bay and the Nehalem River from the Pacific Ocean (Figure 2.3). Since the early nineteenth century, Nehalem Beach has been of archaeological and historical interest due to the remnants of a shipwreck that can be found

FIGURE 2.3. Nehalem Bay, at the mouth of the Nehalem River in Tillamook County, Oregon. (Open access image via U.S. Army Corps of Engineers Digital Visual Library, http://images.usace.army.mil)

scattered along the beach and adjacent area (Erlandson, Losey, and Peterson 2001; Peterson et al. 2011; Williams 2008). For years, archaeologists, historians, and beachcombers have documented and collected the remains of large quantities of beeswax, teak timber, porcelain sherds, and other flotsam and jetsam. Native American oral histories, which stretch back much further, mention a wrecked vessel (or multiple vessels) at Nehalem Beach where abundant beeswax could be found and traded before and after Euro-American settlement. The first written account of the wreck comes in 1813 from the fur trader Alexander Henry, who mentions beeswax brought to Astoria, Oregon, for trade by local Indians (Peterson et al. 2011). Throughout the nineteenth century and into the twentieth, newspapers in Oregon, California, and throughout the country occasionally reported on the "Beeswax Wreck," describing intact remnants of the wreckage and speculating on the origin and history of the vessel.

Menzies claims that the wreck was one of the Chinese junks in Zhou Man's armada, blown off course as the larger fleet explored the western shores of North and South America from Canada's Vancouver Island south to Ecuador. Menzies (2002:239) writes: "Several very old wrecks have been discovered there on the beach at Neahkahnie. One was of teak with a pulley for hoisting sails made of calophyllum, a wood unique to south-east Asia. The wood has yet to be carbon-dated, but if it proves to be from the early fifteenth century it will provide strong circumstantial evidence that one of Zhou Man's junks was wrecked off Neahkahnie Beach. Some examiners of the wreckage there claim to have found paraffin wax, which was used by Zheng He's fleet to desalinate sea-water."

In the summer of 2007, an interdisciplinary team of scientists formed "The Beeswax Wreck Project Group" to locate intact portions of the wreckage; synthesize what is known about the artifacts, features, and history of the material remains; and decipher its likely origins. Washington State Department of Archaeology and Historic Preservation archaeologist Scott Williams (2007) and his team made a number of remarkable discoveries concerning the possible chronology and origin of the wreckage. Reanalysis of all the recovered porcelain sherds suggests that they were manufactured in China during the Kangxi period between AD 1650 and 1700. The only metal artifact that is definitely associated with the wreck is a small silver oil jar, recovered in 1898. Based on stylistic comparisons, the jar was likely produced in the seventeenth century or slightly later and was used in Catholic ceremonies. The first radiocarbon dates were run in the 1960s and 1980s on beeswax samples and yielded age ranges between AD 1485 and 1800, with the most likely overlaps falling between AD 1600 and 1700. For a more recent beeswax sample submitted in 2008, accelerator mass spectrometry (AMS), a high-precision radiocarbon dating technique, yielded an age range between AD 1660 and 1800 (Peterson et al. 2011).

Menzies fails to mention these data, some of which, admittedly, were published after the release of *1421*, but he also never explores in his book the long history of trans-Pacific commerce between Mexico and the Far East beginning in the sixteenth century and ending in the early nineteenth. In a pattern firmly established by AD 1572, Spanish trading ships sailed once or twice a year from Manila in the Philippines, across the Pacific, to Acapulco in Mexico. These long and dangerous voyages were the economic lifeblood of the Spanish colonial empire. Eastbound galleons, some weighing 2,000 tons, built of Philippine hardwoods and capable of carrying a thousand passengers and hundreds of tons of cargo, exchanged Chinese silks, spices, and porcelains and Asian beeswax for Mexican silver. Westbound ships would return to Manila and use their profits to buy additional Chinese goods and provisions for a return trip to Acapulco.

The voyage east was especially dangerous. After departing Manila, the galleons would head north to Japan, where they caught the westerly trade winds. The northern California and Oregon coasts are notoriously foggy, rugged, and stormy, so the galleons aimed for southern California before turning southward to follow the coastline down to Acapulco. Some of the ships strayed off course, met bad weather, were overrun by pirates, or ran aground. When possible, the galleons were salvaged, but some disappeared without a trace.

Based on archaeological and geological fieldwork in 2007, extensive archival research, and analysis of the artifactual remains, the Beeswax Wreck Project Group concluded that the Beeswax Wreck at Nehalem Bay is the remains of a marooned Manila galleon. They have even narrowed the likely candidates to two named vessels in the archival records, the *Santo Cristo de Burgos* of 1693 or the *San Francisco Xavier* of 1705. It is hard to dispute these findings, especially since many of the beeswax blocks have the initials "IHS" or symbols carved into them that are typical of Spanish shipping marks. As the project team continues its work in the coming years, they may be able to tell us even more about the history of the Beeswax Wreck, but one thing is for certain, it is not the remnants of a lost junk from Admiral Zheng He's fleet.

A Fifteenth-Century Chinese Junk in Sacramento?

Menzies's second major piece of physical evidence for precolumbian Chinese along the Pacific Coast of North America comes from the Sacramento River. Menzies (2002:241) describes becoming frustrated with the lack of physical evidence along the western shore of North and South America. In an attempt to spread his message and appeal to the public for help, Menzies gave a lecture at the Royal Geographical Society of London in 2002. As he (2002:241) describes, this put him on the trail of the Sacramento junk:

> It was broadcast around the world; within forty-eight hours reports began to come in from California, drawing my attention to the wreck of a medieval Chinese junk buried under a sandbank in the Sacramento River off the north-east corner of San Francisco Bay.... As soon as I had carried out some preliminary research, I discovered that the prevailing north-easterly winds on the coast

> could have blown the junk straight across the bay and into the Sacramento River.... It was indeed possible, if not probable, that a junk entering San Francisco Bay would have been driven by the winds into the Sacramento River. The tide would also have carried the ship upstream as far as Sacramento.

Menzies (2002:243) includes a map of the San Francisco Bay and a series of arrows depicting the prevailing winds. Setting aside a number of anomalies in his map, including an arrow pointing up the San Joaquin River with a label that suggests it is the Sacramento River, the actual distance between the supposed buried junk and the Pacific Ocean is closer to 200 than 100 miles. Moreover, there are numerous Pleistocene gravel deposits, shallow water depths, sharp oxbows, and fast-moving westerly currents that would have made such a journey by a Chinese junk in the early fifteenth century highly unlikely.

The legend of the buried Chinese junk dates back to the mid-1930s, when two farmers drilled a well in an orange grove near the town of Glenn, California. The pair struck something dense and pulled out a fragment of hard, gray metal. Mystified, they showed it to the principal of a local elementary school, who brought the strange item to an anonymous expert in Red Bluff, California, who determined that it was from a Chinese suit of armor. From there, the story grew into the legend of an ancient Chinese boat, buried under 30 to 40 feet of sediment and gravel in an old channel of the Sacramento River. It became known as the "Glenn Junk," despite the fact that no one had ever seen a vessel and the metal object went missing (Wiley 1975, 2001).

Intrigued, Dave Stewart, a local farmer, began a hand excavation to search for the junk in 1980. After an enormous amount of time and effort, his search ended at 24 feet below the surface when he dipped below the water table and the bottom of his pit flooded. It took nearly 20 years and the encouragement of Dr. John Furry, owner of a local private museum, to revive the search. Together they conducted further explorations, drilling small auger test holes at the purported site location. One hole yielded what Stewart and Furry claim are pottery fragments and millions of tiny black seeds, which, according to them, are the buried contents of a Chinese jar. They also claim to have radiocarbon-dated wood fragments that place the age of the find within Menzies's precolumbian timeframe (Wiley 2001).

Despite the issues of geography and geology, the major problem with the Glenn Junk is that none of the claims have been independently verified. The radiocarbon dates have never been properly reported or shown to have any association with a buried Chinese vessel. The jar fragments have never been identified as pottery sherds, and the seed remains are still unidentified. All of this research, then, is nothing more than the legend itself, wild speculation by someone interested only in providing "evidence" that supports his claims.

In a 2004 *Sacramento Bee* article, it was announced that Menzies and his team of researchers had called off the hunt for the Glenn Junk (Wiley 2004). Menzies claimed that funding would be better spent elsewhere and that the lack of cooperation from local landowners made continued work impossible. What he failed to mention is that an independent team of archaeologists from the University of Denver had conducted electromagnetic conductivity surveys in 2003 and found no evidence of a buried vessel. Rather, their (lack of) findings caused the archaeologists to cancel plans for augering and full-scale excavations.

The final piece of physical evidence of precolumbian Chinese along the North American coast comes from Los Angeles and the purported recovery of ancient Chinese junk anchors. Curiously, Menzies (2002:274, 451) never describes these anchors, except to list them on maps documenting evidence of visits by Chinese treasure fleets to the Americas and of Chinese bases throughout the Pacific Ocean. It's not very difficult to decode the anchors that Menzies is referring to here, especially if you live in southern California and are interested in maritime history and archaeology.

Fifteenth-Century Chinese Anchors in Los Angeles?

The first appearance of ancient Chinese anchor stones in southern California came in 1973 when a U.S. Geological Survey dredging vessel working off the Palos Verde peninsula in Los Angeles County brought up a carved donut stone (Pierson and Moriarty 1980, 1981). Two years later, two local sport divers, Wayne Baldwin and Bob Meistrell, recovered approximately 20 additional carved stones, many with a perforated center for (presumably) the threading of an anchor rope, in a location around a reef at about five meters deep (Figure 2.4). At the time of their recovery, the finds generated a firestorm of publicity. Was this evidence for precolumbian

FIGURE 2.4. Two large "Chinese anchor stones" that now decorate the outside of Bob Meistrell's home in Redondo Beach, California. (Photo by author)

Chinese mariners in southern California? Since Menzies's tale would not be published for over 20 years, people scrambled to explain the origin of the curious stones.

The possibility that captured the most attention was the Chinese legend of the land of Fusang. According to dynastic records, 1,500 years ago a Buddhist monk, Huishen, sailed to a distant eastern land with a group of colonists, never to return (Feder 2008). Depending on whom you ask, the story of Fusang and the Buddhist priest is fact, fiction, or something in between. Scholars argue about the veracity of the story and suggest various localities in the New World as a possible landfall, but the most likely location, if the account is true, is Sakhalin, an island that is part of Russia and just north of Japan. The existence of Fusang has never been verified, and scientific evidence is lacking; still, the anchor stones of Los Angeles County once offered evidence for the Fusang "saga" and, now, the Menzies tale.

In 1980, James Moriarty III, professor of history and archaeology at the University of San Diego, and his colleague Larry Pierson gave credence to the stones, arguing in two academic publications, the *Anthropological Journal of Canada* and the *Proceedings of the Twelfth Conference on Underwater Archaeology*, that they were identical to anchor stones used by Chinese sailing vessels in the Orient as far back as AD 500 (Pierson and Moriarty

1980, 1981). Based on the massive size of at least one of the anchors, they determined that it was carried by a sailing ship measuring at least 100 feet. In addition, they claimed that the objects were made of a sandstone found only in southern China.

In a more balanced piece in the popular publication *Archaeology* in 1982, Frank J. Frost, professor emeritus at the Department of History, University of California–Santa Barbara, noted that the rocks were examined by the UC Santa Barbara Geology Department and determined to be of local origin, most likely Monterey shale. If the objects were constructed from parent material found only in China, the case for trans-Pacific contacts would be much stronger. Although this origin doesn't disprove Menzies's hypothesis, it doesn't support it either.

There are two major problems with using stone artifacts collected by dredging the ocean floor or by amateur collectors with scuba gear to tell an archaeological story. First, there is no effective way to date the anchors: radiocarbon dating will not work (this is effective only on organic material), and there are few other options outside of stylistic dating. Unfortunately, the expedient and utilitarian carving of a hole or holes in stone is not temporally diagnostic. Second, the most valuable part of any archaeological material is its context or association with everything else around it. Imagine finding a broken piece of pottery in a shoebox in your attic, collected years ago by a relative. Without knowing exactly where that pottery sherd came from and what was found around it, an archaeologist would have a very hard time providing much information about the people who made it. Such is the case with the Palos Verdes stone anchors. Artifacts lacking contextual information have very limited explanatory value. This is one of the primary reasons archaeologists spend so much time educating the public about looting and asking hikers not to pick up artifacts when they happen across them.

We can look to the historical record for other explanations, however, which may be more probable than precolumbian contacts between China and the New World. One is that the stones are, indeed, Chinese junk anchors. As we explore in the coming chapters, Chinese fishermen commonly traversed the California shore in locally manufactured junks and sampans during the nineteenth century (Van Tilburg 2007). These anchors (or net weights) may be, then, debris from the early Chinese fishing industry in coastal California. Another possibility is that they were used during historical times, again in the late nineteenth century, by Portuguese whalers. Palos Verdes, once known as Portuguese Bend, was the location of a large whaling station. In fact, Google Maps includes "Portuguese Bend" as a place-name on modern maps of the peninsula. The anchors may have been used to moor whaling boats or the whale carcasses themselves during butchering. Without additional evidence for precolumbian Chinese artifacts, both of these scenarios are much more likely than precolumbian Chinese voyages.

Precolumbian Vikings as a Template

Menzies's evidence for Chinese exploration and colonization of the globe between AD 1421 and 1423 stands in stark contrast to that for the Norse settlement of North America some four centuries earlier. Unlike the case of the proposed Chinese contact, Leif Erikson and his countrymen left a diverse array of quintessentially Viking material culture, despite a relatively brief occupation and a deliberate evacuation that would have included the collection of much of their belongings. Regardless, archaeologists have identified Viking iron, pins, spindle whorls, and house foundations at L'Anse aux Meadows with clear archaeological context and falling within the correct timeframe. Archaeologists have stopped short of definitively linking L'Anse aux Meadows with the settlement described in the sagas, but such a connection may never be possible. For all of Menzies's insistence that his story of fifteenth-century Chinese global exploration is supported by irrefutable scientific evidence, the reality is that not a single piece of Chinese material culture has been recovered from clear archaeological context in the New World, Australia, or any other location that was supposedly visited by Admiral Zheng He's fleet. This is strange considering that the treasure ships and junks (according to Menzies) were loaded with exotic animals, glass beads, Chinese porcelain, and other clearly Old World, Asiatic items.

Despite the poor science, Menzies continues to churn out (and sell) new editions of *1421*, and the public is apparently hungry for more. In 2008, Menzies released his second book, *1434: The Year a Magnificent Chinese Fleet Sailed to Italy and Ignited the Renaissance*, another piece of pseudoscience and pseudohistory that has been maligned by reputable historians from universities around the world. In his third book, released in 2011 and titled *The Lost Empire of Atlantis: History's Greatest Mystery Revealed*, Menzies strays even further into the lunatic fringe. In it, he links the legendary lost continent of Atlantis to the Minoan civilization and claims that it controlled a maritime empire stretching from India to the American Great Lakes.

Perhaps most disturbing is that Menzies envisions his work as an ongoing research project despite the lack of supporting evidence for his claims. He employs a team of researchers and encourages the public to send him new findings that are integrated into future editions of *1421* and included on his website (www.1421.tv). PBS, which, like NPR, should adhere to the highest standards of journalism, even made the book into a documentary. In essence, Menzies is teaching his readers to study history in the same slipshod fashion he does—with little to no training, without a scientific toolkit, and with selective observation. What Menzies doesn't do and what he doesn't teach is how to be a good skeptic. He doesn't create multiple working hypotheses, consider all knowledgeable points of view, independently confirm the "facts," look for weak links in the chain of argument, or see if a hypothesis can be proven false. Menzies has created a historical fact and set out on a mission to prove his ideas, in part by asking the public to send him selective fodder for his book and website—anything that seems to fit, then, does.

Science in the Modern World

The problem is not limited to precolumbian Chinese and Gavin Menzies but is part of a growing trend. That is, despite the fact that we have more techniques and tools than ever before to evaluate claims in science and history, there has been a surge in the spread of misinformation cleverly disguised to look like fact (Shermer 1997). The problem seems to be a lessening of the standards of evidence and a willingness to believe anything so long as it captures our imagination or heartfelt desires. The journalist and author Damian Thompson (2008:12) calls this "counterknowledge": "One of the distinguishing features of counterknowledge is a casual approach to the truth: it encourages and takes advantage of a significant lowering of the standards of proof in society generally. As a result, we are exposed to all sorts of empirically dubious information, much of it conceived in a spirit of credulity or carelessness rather than actual deceit." PBS and NPR fell victim with Gavin Menzies and *1421* and fans of the History Channel's "Ancient Aliens," and scores of other such television shows, documentaries, and conspiracy theories do as well. Whenever I lecture on the ancient Maya of Mesoamerica in my "Introduction to Archaeology and World Prehistory" university class, for example, I am inevitably asked about evidence of contact with aliens or ancient astronauts.

The question remains, what do we do about this? There is no easy answer, but I believe it is, partially, the job of academics to respond to counterknowledge, pseudohistory, and pseudoscience and to challenge those who choose not to play by the rules of science. Too often, we have ignored the outlandish in hopes that it would go away or have concentrated on scholarly endeavors that garner accolades, such as grant funding, course releases, and tenure and promotion at colleges and universities. Part of any academic's job should be to disseminate his or her work to the public in clear and interesting ways and respond to those who profit from the spread of misinformation, such as Gavin Menzies. Granted, such work is not very profitable or popular. For example, arguably one of the most successful pseudoscientific books in history, Erich von Däniken's *Chariots of the Gods?*, was an international best seller, was translated into dozens of languages, and sold by the millions, but the scientific response, Clifford Wilson's *Crash Go the Chariots* (although successful and a best seller), reached only a small fraction of that readership. To that point, it is critical that members of the public ask questions about the "facts" they hear, read, and are told. Often the answers are right at our fingertips if we choose to look, using logic, simple statistics, or even a Google Scholar search. On the other hand, being a positive skeptic can be arduous, difficult, painstaking work, but the rewards are a deeper understanding of the way the world is, was, or will be.

Fortunately, there are plenty of real histories that can capture the attention and imagination of the public. Nineteenth-century Chinese immigration to the California shore and the genesis of commercial abalone fishing is just one of these. Armed with a better understanding of the scientific method, let's return to that story and fast-forward 400 years from Menzies and Admiral Zheng He to imperial China and the waning years of the Qing dynasty. It is this moment in Chinese history that sets the stage for immigration to California. Before we can understand why it was nineteenth-century Chinese immigrants who founded California's commercial abalone fishery, we must understand why they left their families, livelihoods, and homes for the uncertainty of a land and a life half a world away.

CHAPTER 3

Chinese History and the Exodus to Gold Mountain

Mid-nineteenth-century China was a land of mystery and intrigue for many Westerners, and perhaps not much has changed. Even today, China can be an enigmatic place, half a world away and so vast and diverse that it defies description. Covering nearly 10 million km^2 and inhabited by 1.3 billion people, China is the most populous nation on earth (Figure 3.1). Although only slightly larger in land area than the United States (and slightly smaller than Canada), China has an astounding ecological diversity within its borders. China's northern and western margins are defined by the Siberian mountains of Russia and the Indian Himalayas, and to the east by the Bohai, Yellow, East China, and South China Seas, totaling 14,500 km of Pacific Ocean coastline. The isolated and rugged plateaus of southwestern China are bordered by temperate grasslands and forest steppes to the north and northwest, along with the harsh, arid sands of the Gobi and Talkimakan Deserts. Lush subtropical forests are found near the damp frontier to the southeast, along China's borders with Laos and Vietnam. A number of important river systems originate in the Tibetan Plateau and drain to the east; the two largest, the Yangtze and Huang He (Yellow) Rivers, wind their way to the densely populated eastern seaboard.

Along with its varied landscape, China possesses rich linguistic, cultural, and ethnic diversity. Nearly 300 languages are spoken in China, spanning eight language families, most with considerable morphological and phonetic differences (Kane 2006; Ramsey 1987). A myriad of cultural and ethnic groups live in China and maintain diverse dress, foods, music, and customs. Officially, however, China is composed of 56 ethnic groups,

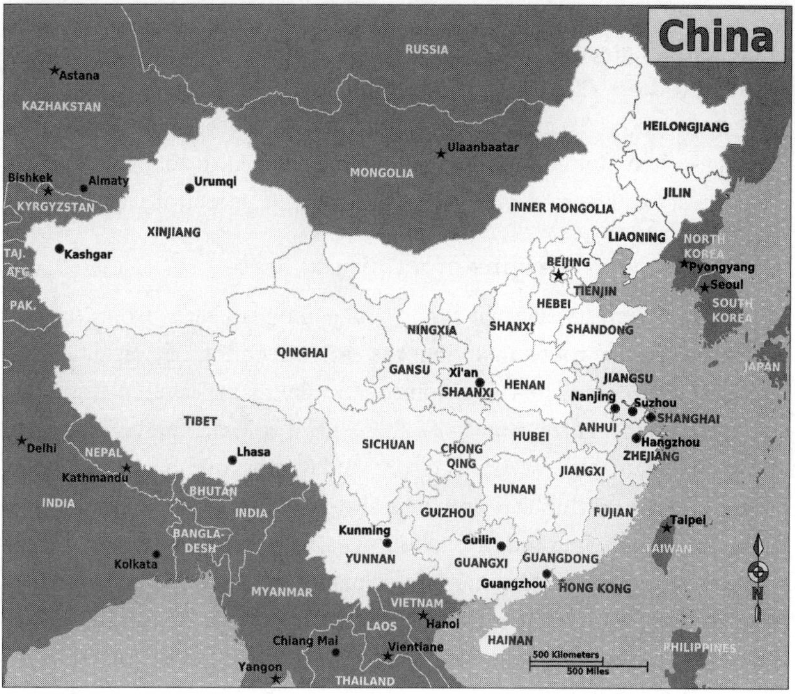

FIGURE 3.1. Modern geopolitical map of China, its provinces, and surrounding countries. (Open access image via http://www.sacu.org/provmap.html)

the dominant Han and 55 minorities. These minority groups were identified by a team of social scientists (ethnologists and linguists) and Communist Party cadres, beginning in the early 1950s, shortly after Mao came to power, as part of the Ethnic Classification Project (Mullaney 2010). Driven by politics, the classification system is deeply flawed because many of the official ethnic groups of the People's Republic of China (PRC) contain, within themselves, diverse clusters of people speaking mutually unintelligible languages and following very different cultural norms.

From this diversity, both geographical and cultural, developed one of the world's first nation-states. Before we can understand how it was that Chinese immigrant laborers and fishermen, most from a single province over 11,000 km or nearly 7,000 miles away, constituted the single largest ethnic group to flood California during the Gold Rush, we must first trace the rise of the Chinese state. When and where did the foundations for complexity and state-building in China begin? What were the pivotal moments in Chinese history that resulted in the rise and fall of dynastic lineages? What governmental and social policies led to the mass exodus in the mid-nineteenth century? And what confluence of factors uniquely positioned these Chinese immigrants to found California's commercial abalone industry? To answer these and other questions, we begin 15,000 years ago when hunter-gatherers populated China at the dawn of some of the earliest experiments with domestication and agriculture in the world.

Origins of a Chinese State

China's real majesty, at least for an archaeologist, stems from its deep cultural historical roots, including the early invention of domestication and agriculture and the transition to complex, state-level, hierarchical sociopolitical organization. Given the central role rice plays in the diet and cuisine of many East Asian cultures today, most of the Western world probably assumes that this crop was always the most important domesticated plant of China and throughout Asia. In reality, however, there are over 280 separate taxa of domesticated plants and animals known to have been used in East Asia. Determining which taxa were the earliest and why they were domesticated is often a considerable challenge for archaeologists and paleobotanists.

Domestication is defined as changes in the physical characteristics of a plant or animal species caused by human manipulation: humans control the breeding of animals or the sowing of seeds in order to select for desirable traits such as docile dispositions or large seeds and fruits. These changes take place gradually and are better thought of as processes than events. The "Agricultural Revolution," as it has often been called in archaeological and popular literature, was a revolution in slow motion—playing out over millennia. Defining an absolute break between foragers and food producers is impossible. Once humans start to select wild seeds that are larger, for example, the question becomes, when are these seeds large enough to be considered domesticated? Morphological changes do not take place overnight, and the transition from wild progenitor to domesticated plant is a gradual one.

Reliance on domesticated plants and animals as dietary staples is also gradual and part of a cultural continuum that can evolve over thousands of years. The Smithsonian archaeologist Bruce Smith (2001) called this "low-level food production," in which hunter-gatherer populations gradually incorporate domesticated plants and animals into their diets as they transition from hunter-gatherer-foragers reliant on wild food sources to full-scale agriculturists dependent on domesticated plants and animals.

The other, perhaps less academic, problem in designating when cultures domesticated plants and animals is largely a taphonomic one. Taphonomy is the study of how the archaeological record forms. This includes all the processes that affect animal and plant remains, artifacts, and features after they are deposited in the ground. Unlike chipped and ground stone tools and other more robust parts of the archaeological record, seeds, nutshells, and bones are much more delicate and are often not preserved, especially in damp environments or acidic soils. Marking the invention of domestication and the transition to agricultural societies in any region, then, is often a difficult task, and chronologies are constantly in flux as new sites and better data become available.

Domestication and agriculture in China began in much the same way as in other regions around the world, although these processes started earlier in northern China than many other places. Even during the height of the Last Glacial Maximum (LGM), between 23,000 and 19,500 years ago, hunter-gatherers along the Great Bend of the Yellow River began experimenting with local plants. At Locality 14 of the Shizitan site in North China's Loess Plateau, the Stanford University archaeologist Li Liu and colleagues conducted usewear and residue analyses on three grinding stones (Liu et al. 2013). Starch residues from wild grass seeds, beans, yams, and snakegourd roots were all identified, suggesting that hunter-gatherers were becoming more reliant on wild plant foods during this harsh period of cold global climates and resource scarcity. This early experimentation in northern China may have given local Paleolithic foragers a 10,000-year head start in learning about the unique properties of these plants, which ultimately led to very early domestication in the area (Liu et al. 2013:5385)

After the end of the LGM, about 18,000 years ago, worldwide environments rapidly changed, marked by a distinct warming trend, the retreat of polar ice and glaciers, and the rise of eustatic (global) sea levels. At the same time as many large Pleistocene animals (megafauna such as mammoths and mastodons) went extinct, humans were forced to adopt new behaviors and strategies to deal with shifting environmental conditions. These changes were more extreme in some places than others and occurred at different times in different places, but a number of commonalities have been identified worldwide: the diversification of human diets to include a greater reliance on hunting small animals, fishing, and collecting plants; new stone tool technologies; specialized hunting strategies; population growth, regionalization, and increased sedentism; the exchange of exotic items such as metals, stone, and shell; and the foundation of social differences between individuals.

In China, the environmental changes of the Mesolithic (the name given to the transitional period spanning the end of the last Ice Age to the rise of agricultural societies), from about 15,000 to 10,000 years ago, were more extreme in the north than the south. In northern China, warming temperatures and rising seas allowed southerly, warm-weather-adapted terrestrial plants and animals to expand their ranges into mid- and northern latitudes (Higham 2005). Thousands of kilometers of new bays, marshes, and estuaries were created, along with midlatitude woodlands and grasslands. Under these resource-rich conditions, Mesolithic cultures in northern China developed a new blade-and-flake industry along temperate forests and a microblade industry in riverine and lake habitats.

Chipped stone blades mark a technological innovation that uses less raw material (valuable silica-rich stones such as chert, quartz, or obsidian) to produce more cutting edge than previous lithic technology. These blades, generally less than 50 mm long, are at least twice as long as they are wide and can be hafted along the margins of bone, antler, or wood shafts for harvesting plants or used as the tips to barbed spears as part of a deadly hunting technology. Armed with this new technology, hunters sought a variety of large game such as elk and ostrich and small game such as badger and fox in forests and grasslands. Fish became the staple resource along productive freshwater lakes.

In southern China, where environmental change was not as extreme, a new set of Mesolithic adaptations also took shape, including the invention of a chipped pebble stone technology called Hoabinhian (Feder 2011: 246–247). Microanalysis of edge damage suggests that many of these stone tools were used to fashion wooden implements, but unfortunately, we have very few preserved examples of these wooden tools which might tell us about the larger Hoabinhian toolkit. What we do know is that these cultures focused on wild cattle, small game, shellfish, and wild plants to furnish a diverse subsistence economy.

Archaeologists long believed that the epicenter of early domestication was the Fertile Crescent—the rich agricultural lands surrounding the Tigris and Euphrates Rivers—in modern-day Iraq and Iran. It is here that archaeologists dated domesticated barley kernels and emmer and einkorn wheat seeds that are distinctive in size and morphology from their wild cousins to between 10,000 and 11,000 years old. More recently, archaeologists have found evidence that figs may have been domesticated in the area even earlier, at about 11,400 years ago (Kislev et al. 2006). It is only recently, however, that archaeologists have successfully radiocarbon-dated rice grains and husks from the middle Yangtze River in China to 12,000 years old (Crawford and Shen 1998). The University of Cambridge archaeologists Martin Jones and Xinyi Liu (2009) even recovered early ceramic vessels with embedded charred rice remains. Domesticated rice, then, probably originated from wild stocks near swamps on the border of the Himalayan Mountains.

After rice was successfully domesticated, China began a steady march toward the complex state we know today. An international team of scientists, led by Richard MacNeish, has excavated two important cave sites in southern China, Xianrendong and Wangdong, and found evidence that the inhabitants began rice cultivation, the production of farming tools, and the invention of some of the world's earliest pottery more than 11,000 years ago (MacNeish et al. 1998). Near the modern city of Shanghai, south of the Yangzi River mouth and Hangzhou Bay, archaeologists have found the amazingly well preserved remains of a 7,000-year-old settlement (Liu and Chen 2012). Waterlogged soils along a lake edge preserved the remnants of substantial wooden structures on pilings and the scattered remains of wetland rice stalks, seeds, pottery sherds, and wooden artifacts. Careful analysis of the food debris suggests that people were collecting wild rice and tubers and cultivating increasingly domesticated varieties. Excavations yielded evidence of wild water buffalo, pigs, waterfowl, fishes, and shellfish, suggesting a transitional subsistence economy from generalized hunter-forager to rice farmer and pastoralist. Similar processes were playing out in many other areas of southern China between 10,000 and 4,000 years ago as humans modified wetland environments to make them conducive for rice agriculture (see Crawford 2011; Liu et al. 2007; Zhao 2011).

Over 500 km north of the Yangtze and early rice domestication, a second center of independent agriculture arose in the Huang Ho River Valley of northern China (Liu 2004). Here, in a small basin sandwiched between high-elevation woodlands to the west and swampy lowlands to the east, millet was the agricultural staple, likely domesticated more than 8,500 years ago when sedentary farmers, the Peiligang culture, began to congregate along deciduous forests (Higham 2005). Sedentary, agricultural villages became abundant and widespread by 8,000 years ago. Archaeologists have identified the remains of semi-subterranean houses and storage pits, some with traces of domesticated broomcorn millet and rape seeds (Higham 2005:239). The sites are often littered with polished spades, stone axes, pottery fragments fired in kilns, and other vestiges of an agricultural economy and sedentary lifestyle.

Even larger villages took shape less than 2,000 years later in the area when Yangshao cultures constructed planned communities dependent on domesticated foxtail millet, Chinese cabbage, and, to a lesser degree, rice (Liu 2004). Early Yangshao settlements averaged about 10 hectares but grew tenfold or more over the next several thousand years. Growing population densities and increased competition for resources are manifested in the construction of fortified settlements, complete with rammed-earth walls and moats, and the rise of sociopolitical complexity.

As in many areas around the world, the roots of Chinese civilization, complexity, and state-building followed shortly after the food-producing revolution. Domestication and agriculture may have taken thousands of years to fully form and spread, but once they took hold, there was a dramatic shift to social, political, and economic complexity. By 5,000 years ago, China was a patchwork of large and small kingdoms overseen by elite, wealthy chieftains who vied for political control. This marked the birth of Chinese dynastic traditions that would characterize the millennia to follow (Table 3.1).

Complexity and Dynastic Rule

China's first truly urban centers took shape in the north, near the modern city of An-yang, along the Wei and middle Yellow Rivers (Allan 2002). Legend describes the celebrated ruler and warlord Huang Di as founding the first city over 4,600 years ago. Huang Di ruled with an iron fist, and his harsh government would set the tone for generations of warlords to follow. Huang Di and his successors constituted the Shang dynasty and the earliest expression of Chinese civilization (Figure 3.2).

TABLE 3.1. The Major Dynastic Kingdoms of China.

Dynasty	Time Period	Characteristics
Shang	1700–1027 BCE	Period of extensive state building, monumental architecture, and bronze work. Chinese writing system developed.
Zhou	1027–221 BCE	Iron casting is invented, and large-scale irrigation projects are undertaken. Art and literature, including the teachings of Confucius, become widespread.
Qin	221–207 BCE	Qin Shi Huangdi becomes China's first emperor. Law and tax systems are instituted. Writing, weights, and measurements are standardized. Construction of the Great Wall begins.
Han	206 BCE–AD 220	Trade routes to Europe are established along the Silk Road as well as trade to SE Asia. Buddhism is introduced from India.
Sui	AD 580–618	Transportation networks within China expand, including construction of the Grand Canal, linking the Huang and Chang Rivers.
Tang	AD 618–907	Zenith of Silk Road trading.
Song	AD 960–1279	Invention of moving type and paper money. Printing, poetry, and calligraphy are widespread.
Yuan	AD 1279–1368	Grandson of Genghis Khan, Kublai Kan, founds a dynasty, and his court is famously visited and described by Marco Polo.
Ming	AD 1368–1644	Trade flourishes with European merchants. The Forbidden City is built in Beijing.
Qing	AD 1644–1911	Manchu invaders establish a new dynasty. Internal strife causes the final collapse of dynastic rule.

Sources: Based on Chang (1986); Spence (1990); Twitchet and Loewe (1986).

Over the next several millennia, walled towns ruled by a local elite flourished in northern China. Chieftains directed and controlled agricultural production, organized military campaigns, and established a loosely integrated alliance of small kingdoms, constantly at war (Chang 1986; Liu and Chen 2012). Standing royal armies would swell with thousands of conscripts when war broke out, when rebellious rivals offered threats to the throne, or when sacrificial victims were needed for important ceremonies. Massive tombs, requiring the labor of thousands of peasants, were built as monuments to the cities' kings. Extravagant items of wealth and power were interred with the fallen leaders, including elaborately carved jade, bronze, and ceramic artifacts; chariots and sacrificed horses; and the decapitated remains of human servants.

The Shang dynasty propelled China not only into statehood but also into the historical period. The Shang people developed one of the earliest

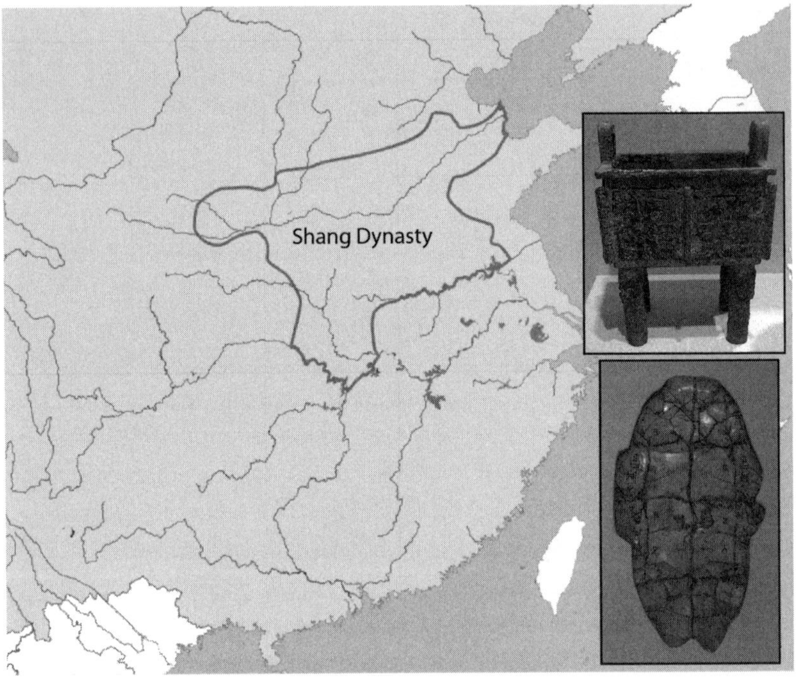

FIGURE 3.2. The extent of the Shang dynasty. *Inset (top)*: Elaborately carved bronze vessel from the Shang dynasty. *Inset (bottom)*: Tortoise shell with divination inscription from the Shang dynasty, stored at the National Museum of China in Beijing. (Composite by author; all images open access via http://commons.wikimedia.org)

writing systems in the world, containing over 5,000 characters, only a fraction of which have been translated (Chang 1986; Liu and Chen 2012). Archaeologists have found the majority of Shang text inscribed on oracle or divination bones. Ancient Chinese priests used tortoise shells and cattle bones to ask a question of the gods. The bone or shell was then heated until it cracked. Based on the patterning of the crack, as read by the priests, the questions and answers were inscribed on the surface of the bone or shell (see Figure 3.2, bottom inset), giving archaeologists fascinating insight into royal policies and history. Much of the text covers war, hunting, agricultural success, and the wealth of the elite. Human or animal sacrifices were often made to divine ancestors, who could provide favorable future outcomes (Kane 2006).

Equally impressive is the expert bronze work of the Shang dynasty (see Figure 3.2 inset photograph at top). While many of the bronze artifacts found at Shang sites are utilitarian tools, including food and drinking vessels, knives, chisels, axes, adzes, and other weapons, the most stunning pieces are ceremonial objects unearthed from royal tombs (Keightley 2000). For over five centuries, Shang bronzesmiths perfected their sophisticated and elegant artwork. The most elaborate pieces were careful copies of clay prototypes, fashioned around a clay core and encased in a mold. Once the clay version was complete, the outer mold and inner core were removed and the remaining parts assembled to receive molten bronze. The result was a dramatic and complex piece of art, unrivaled in the West.

By about 3,000 years ago, the Shang dynasty was overthrown by a rebellious neighboring group, the Zhou. Rather than institute wholesale change, the conquerors integrated the Shang network of towns and officials into their own state system (Liu 2004; Liu and Chen 2012). This action shifted the center of political and economic power away from the north to the south and west along the Wei and middle Yellow Rivers. The Zhou created a network of nearly autonomous provinces throughout China but, in keeping with the Shang tradition, continued large-scale agricultural production, sanctioned major engineering projects, and maintained a tradition of military conflict (Liu and Chen 2012). Large cities during the Zhou dynasty grew to provincial capitals within a densely populated landscape of farming villages and small towns. The Zhou dynasty also saw the flourishing of literature, art, poetry, music, and trade, making it a culturally significant period for China and Chinese historians.

After nearly a millennium of rule (divided into the Western and Eastern Zhou dynasties), the Zhou dynasty fell and sweeping change spread throughout China. Historical texts recount that in 221 BC the ruler from a relatively small Zhou state north of the modern city of Xi'an conquered his six major rivals, unified the state, and became China's first emperor. Emperor Qin Shihuangdi, and the Qin dynasty, ruled for only 15 years, but his political, economic, and military reforms and monumental construction projects exerted tremendous influence on all subsequent dynasties.

Emperor Qin is probably best known for his role in the construction of the Great Wall of China (Figure 3.3). This massive engineering project, the majority of which was built during the Ming dynasty, was first conceived during the Qin dynasty. By imperial order, new construction began connecting a series of existing fortifications into one continuous wall that would protect the northern border from military incursions and facilitate

FIGURE 3.3. A small section of the Great Wall of China in 1907. (Open access image via http://commons.wikimedia.org)

trade and taxation along the Silk Road (Barfield 1989). The Great Wall would ultimately extend for 5,000 km along an east-west-trending line, and those small sections that have been restored have become iconic of Chinese history and culture today.

Like many of the dynastic leaders before him, the Qin emperor commissioned a mounded tomb as his final resting place. The scope of Emperor Qin Shihuangdi's burial site, at almost half a kilometer square, is astounding. Although the location remains unexcavated, historical accounts suggest that the tomb contains a finely detailed replica of the

FIGURE 3.4. A small portion of the vast Terracotta Army of Emperor Qin Shihuangdi. (Open access image via http://commons.wikimedia.org)

royal palace surrounded by a river of mercury. Two kilometers away, well construction in 1974 resulted in the discovery of a spirit army buried in two massive pits containing some 3,000 life-sized "terracotta warriors" and associated ceramic models of horses, chariots, and weaponry (Figure 3.4). Examples from the excavations have toured the world, and tourists flock to the site to view the remains under the protection of a vast metal hangar. Archaeological research is ongoing, and the full extent of the site is still unknown.

Emperor Qin Shihuangdi's most lasting legacy, however, was his mandate to standardize the Chinese writing system. Although China contained a tremendous diversity of ethnic traditions, languages, and cultural practices, only one written language has emerged. In order to facilitate the centralization of the Chinese state, Emperor Qin instituted an official script of 3,000 characters (Kane 2006). Unlike the letters of Western alphabets, which represent sounds combined into unique words, pictographs became the basis for the modern Chinese vocabulary. Having a set of universal characters, then, allowed the central government to communicate with officials and the educated public from one end of

China to the other. Emperor Qin's writing system did not just extend to text but also standardized measurements, currency, and weights, providing his government with a powerful grip on communications, trade, and agricultural production across the empire.

The succeeding Chinese dynasties operated in a similar fashion as those that came before. An increasingly poor, rural peasant population expanded agricultural outputs and funded the opulence of the wealthy. Dams, canals, immense agricultural fields, and roads facilitated trade, tribute, and control of the vast empire. The Han dynasty (as with the Zhou, broken into two halves), lasting more than four centuries following the Qin dynasty, opened overland trade with the West along the famous Great Silk Road through Central Asia and with the powerful Khmer empire of the Angkor state in Southeast Asia (centered in modern-day Cambodia). The Han also instituted a centralized training institution to educate administration officials and organized the government into a series of bureaucratic ministries (Liu and Chen 2012). The Han dynasty invested in new technologies in agricultural production; used iron to produce a variety of weapons, domestic items, and industrial tools; built roads to facilitate trade; and carried out military campaigns to expand the empire to the northwest and south. The expansion of the Chinese empire during this time is still reflected on geopolitical maps today, and the dynasty's contributions to art, literature, and government have stood the test of time (Twitchet and Loewe 1986).

After the fall of the Han dynasty, subsequent governments faced many of the same challenges as their predecessors. Dynasties were constantly at war or under the threat of war. Their power waxed and waned through time, and military campaigns were numerous (Spence 1990). For thousands of years, the primary threat to dynastic powers came from inside China, and new leaders rose to power within China's borders. The dominant Han ethnic group was able to repel incursions of "barbarians" from the north, partly due to the protection of the Great Wall. But on April 25, AD 1644, during the Ming dynasty, the unthinkable happened: a nomadic group from Manchuria—the Manchus—breached the wall, conquered the Han, seized control of Beijing, and entered the Forbidden City. On the final morning of his rule, the emperor lamented, "My ministers have failed me. As ruler of the country I [must] die for the altars of soil and grain. An empire that [lasted] 277 years: lost in one day" (as cited in Wakeman

1985:265). Three days later a palace servant found the emperor's body beneath a pine tree; he had strangled himself with his sash.

Rule under the Manchus and the Qing Dynasty

The Manchus were effective and swift conquerors. They overthrew Beijing, established their own capital in the city, and occupied the Forbidden City, the home to generations of former emperors and dynastic governments. The new leadership faced a number of immediate and pressing tasks, including restoring administrative order, pacifying the countryside, and quelling remnant military resistance. The longer-term goal was to reform the bureaucratic system and provide economic opportunities to a populace that had become entrenched in poverty and oppression under Ming rule (Wakeman 1985:414). Rather than create a new governmental infrastructure, however, the savvy Manchus exploited the existing system (although they instituted some changes) and established their own dynastic line, the Qing dynasty. The new government was able to win quick (albeit tenuous) support among the masses by promptly introducing public welfare measures and dramatically reducing the heavy taxes imposed by the Ming (Wakeman 1985:449–450).

The Manchus also instituted a number of more insidious decrees as a way of signaling the subjugation of the Chinese Han to Qing rule. They quickly outlawed intermarriage between Han and Manchus (Chang 2003:7) and forbade Manchu women from following the Han custom of foot binding (Spence 1990:39). Although foot binding was extremely painful, it had become a centuries-old symbol of status and beauty, practiced by elite and commoners alike. By refusing to follow this custom, the Manchus asserted their cultural independence and erected another barrier to intermarriage—Han men found the unbound feet of Manchu women to be sexually unattractive (Spence 1990:39).

In July 1645, the Manchu prince Dorgon decreed that every Chinese man must shave his forehead and begin growing his hair in a long braid (Spence 1990:38). To cut one's hair was an act of treason and punishable by death (Chang 2003:7–8). This was the traditional hairstyle—called the queue—among the Manchus; the Han preferred long and elaborately dressed hair as a signal of masculinity and sophistication. Subsequent decrees required the adoption of Manchu-style clothing, a tight jacket with a high collar fastened at the right shoulder, over the loose robes of

Ming tradition. Each of these forced adoptions of Manchu cultural customs was a symbol of Han submission to the new imperial rule of the Qing dynasty.

At a time when the vast majority of China's population lived as poor peasants performing the backbreaking labor of agriculture, the Manchus' most ingenious move was to maintain the imperial examination system (Chang 2003:7–9). The Manchus recognized that peasants labored away in rice plots and agricultural fields; slept on woven mats in rustic houses; wore plain cotton clothes, simple leather or grass sandals, and nondescript broad-brimmed hats; and, from this humble position, served as the nation's economic backbone. These farmers had little connection to their government and even less hope of ever improving their lot in life. Keeping them loyal, or at least placated, was essential to sustaining the privileges of the Manchu elite and ensuring that the wheels of government kept turning. To this end, the Manchus maintained a series of civil service exams on ancient Chinese literature and philosophy, a means by which anyone could elevate their economic and social status through hard work and education—or so it seemed. An assiduous, intelligent peasant farmer, then, could take local, provincial, and national exams to reach the upper echelon of society. This fit in perfectly with the Confucian ethic of hard work, education, and tradition. The reality, however, was that the system was far from equitable. Women were not allowed to take the exams, and sons from wealthy families could afford the best and brightest tutors to coach them (Chang 2003:7–9). The poor faced enormous odds against ever passing the exams and moving through the civil service system. But those who did, the best and brightest of the Han, were quickly integrated into the Manchu system and became representatives of the oppressor rather than agents of change.

Isolation, British Opium, and Civil War

The policy of isolation from the outside world that began during the Ming dynasty continued under Manchu rule. For a brief time in the early fifteenth century, China was a maritime power that explored the eastern world from the South Seas to the Indian Ocean. Ming emperors dispatched a series of maritime explorations, which increased foreign trade and tribute as far as the eastern shore of Africa (Duyvendak 1939). The most extraordinary of these ventures were the naval explorations of the eunuch admiral Zheng He (see Chapter 2). Zheng He commanded seven expeditions between about 1405 and 1435 with a mission that was largely political but also designed to stimulate trade with other Asian nations (Zhenghai and Chuankang 1996). The earliest of these explorations included 62 ships, 28,000 men, and cargo holds packed with Chinese silks and porcelains. Zheng He's fleet returned with exotic foreign products such as spices, precious metals and stones, ivory, animal hides, feathers, and tropical hardwoods, along with many foreign ambassadors who declared their nations tributaries of the Chinese empire (Dreyer 2007). These famous "treasure fleets" were commissioned by the Ming emperor Yongle but lasted a mere 30 years. After Yongle's death while campaigning against Mongol invaders in 1424, one final symbolic expedition was launched, after which succeeding emperors Hongxi and Xuande allowed Zheng He's massive fleet to rot in port (Dreyer 2007; Ma 1970).

The abrupt end to these naval excursions was part of a growing xenophobia in China, fueled by Confucian scholar-bureaucrats who advocated a policy of isolation and a self-sustaining agrarian system. After Emperor Yongle's death, attention returned to the homeland borders, in no small part due to the need for money to support national defense against Mongol attacks.

Zheng He's trade and tribute missions were enormously expensive. As many as 2,000 ships were constructed for the seven maritime treasure voyages (Dreyer 2007). Money that once supported exploration now subsidized attacks against the Mongols and construction of the Great Wall. The state also believed that overseas trade would fuel unrest and turmoil. Government officials were concerned that secret information about national defenses would be leaked to foreign powers and encourage piracy and other crimes (Spence 1990:57). Confucian scholars so adamantly opposed Zheng He's voyages that, in the late 1470s, an official burned most of the court records related to them, and any mention of the expeditions was suppressed (Dreyer 2007; Finlay 2004:232–233). By the end of the fifteenth century, the dynasty codified its policy of isolation and imperial subjects were forbidden to build oceangoing ships or to leave the country.

The Qing dynasty of the Manchu conquerors maintained the Ming policy of isolation, although perhaps for slightly different reasons. The Chinese elite viewed their country as the center of the civilized world, and all other countries were peripheral (Spence 1990:119). The Chinese showed little interest in the languages, customs, and geography of the outside world. Although trade with other nations occurred, it was tightly regulated and the government was largely uninterested in its economic potential. Officials went so far as to regulate not only the location and frequency of foreign trade but also minor details such as the types of goods and personnel involved (Spence 1990:119–120).

The Qing government also feared that overseas Chinese emigrants would ally themselves with rebel forces in Taiwan and plot the overthrow of the government (Chang 2003:12). The penalty for emigration, then, was death by beheading. The major problem with such a policy is that once the crime is committed, the perpetrator is outside the reach of governmental authority. The law therefore allowed for the execution of those who assisted individuals in leaving the country and provided bureaucrats with merit points, which could result in promotions, for capturing individuals attempting to flee China (Chang 2003:12).

Based on these harsh and isolationist policies, a Manchu government built on Ming foundations, and a set of clever social controls, the Qing dynasty ruled for over two and a half centuries, from 1644 until 1912. Despite all its efforts, however, the Qing grip on the country remained slippery, even after the Rebellion of the Three Feudatories, an uprising of southern provinces (Yunnan, Guangdong, and Fujian) under Han control lasting from 1673 to 1681, was quelled and the Manchu conquest was complete (Wakeman 1985:1099–1105). The Qing government regularly sought out dissidents and killed rebels with the unwavering goal of maintaining power. To this end, the Manchus initiated a system of secret communications between the central government and provincial officials (Fairbank 1986:19). This allowed for greater control of the distant countryside and created a network to funnel secret revenue to the imperial elite. The continental divide between the rich and poor, the haves and have-nots, however, left the minority Manchu ruling class in constant fear of rebellion.

Much of the problem stemmed from the unbridled extravagance and mismanagement displayed by the Manchu elite. Qing officials and bureaucrats lived off the grueling labor of conquered countryside Han villagers (Fairbank 1986:19) by instituting an inequitable taxation system that favored the elite and bled the poor. The isolationist policy, along with bureaucratic apathy, caused widespread economic stagnation. Other than retaining flood-control measures for agricultural fields and copper mining for coin production (Fairbank 1986:19), government officials were uninterested in economic development that could have improved the lives of village farmers and the country at large. Instead, the Manchus developed an elite quasi-welfare state for their own people. For example, the Qing granted military stipends for every Manchu boy at birth (Chang 2003:13). The original intent was to ensure future service in the military, but the system quickly de-evolved into an entitlement scheme for Manchu men, regardless of future military service.

The unfortunate reality for the typical Han laborer or farmer was that the corrupt rule of the Manchus saw little powerful, organized opposition for centuries, and the Manchu elite settled into a comfortable habit of living far beyond their means. The best-selling author Iris Chang (2003:13; also see Gernet 1982:530–531) described the scope of Manchu extravagance: "During the eighteenth and nineteenth centuries, gross mismanagement of state funds almost emptied the coffers of the Qing treasury. In 1735, when Qian Long became emperor of the Qing dynasty, the imperial government owned some 60 million *liang* of silver; subsequent excessive spending sent China on such a downward spiral that by 1850, 115 years later, the reserves had dropped to only nine million *liang*."

An exploding population only exacerbated the problem. Mark Elvin (1993), professor emeritus of Chinese history at Australian National University, estimated that China's population reached 50 million by AD 1 and grew to 100 million by AD 1000, to 200 million by the early 1700s, and to 400 million by the mid-nineteenth century. Compare that to the approximately 20 million people who lived in the United States in 1850—and remember, both countries have about the same land area!

Overcrowding resulted in shortages of farmland, inflated rents for tenant farmers, and concentrated wealth among elite landowners. China tried to curb these problems by importing food from other countries, but desperate villagers felled entire forests and expanded agricultural plots to steep, terraced hillsides. The result was massive soil erosion, declining crop yields, flooding, and widespread famine and starvation (Chang 2003:14).

These troubles were exacerbated by growing tensions with Western imperial powers, who chafed at restrictions on foreign trade. By the seventeenth century, the overseas empires of Spain and Portugal were in decline and Britain, France, and Holland were emerging as global colonial powers. In particular, the British East India Company, founded in 1600, was quickly rising from a small operation to a major global trading outfit (Wills 1974). The British government granted it a monopoly on all trade in east India, and by the mid-eighteenth century the company was actively trying to increase its presence and influence in China (Wills 1974). The Chinese government, however, treated foreign merchants with contempt, dictating where they were allowed to live and under what conditions they could conduct business (Chang 2003:14). In major ports such as Canton (Guangzhou), corruption and administrative red tape regularly hindered trade and the basic repair of merchant ships while in port (Spence 1990: 120–121). Stories of supposed maltreatment circulated widely among European nations, growing into anti-Chinese sentiment in Britain and other Western nations.

The East India Company and other foreign traders pushed back, and by the mid-eighteenth century, formal complaints were issued to Peking (Beijing). The Qing response was to restrict all trade to a single port, Canton in the southern province of Guangdong, and forbid all foreign residence outside the trading season between October and March (Spence 1990:121). At the same time, the Chinese tea market was exploding worldwide. The East India Company alone was buying over 23 million pounds of tea annually by 1800, which does not include tea sold to other European and American traders (Spence 1990:122). The British and other European powers could find no Western exports to match the global demand for not only Chinese tea but also porcelain, silks, and decorative goods. As payment, Westerners were forced to trade massive amounts of silver, draining wealth from Western nations and leading to staggering prosperity for the Qing elite.

The British East India Company, desperate to loosen trade restrictions, entered an agreement with King George III's government to send a paid emissary to China, Lord George Macartney (Cranmer-Byng 1963). A seasoned foreign diplomat, Lord Macartney and a large supporting cast of colleagues were tasked with securing British rights for ambassadorial residence in Peking, the termination of the restrictive Canton trading system, the opening of new ports for international commerce, and the establishment of fair tariffs (Cranmer-Byng 1963:3–17). The emperor would not capitulate on any request, instead writing: "The Celestial Empire, ruling all within the four seas, simply concentrates on carrying out the affairs of Government properly, and does not value rare and precious things...we have never valued ingenious articles, nor do we have the slightest need of your country's manufactures. Therefore, O King, as regards your request to send someone to remain at the capital, while it is not in harmony with the regulations of the Celestial Empire we also feel very much that it is of no advantage to your country" (Cranmer-Byng 1963:340).

The British solution to this stalemate came not by any legal means but through the illicit smuggling of opium, a dangerous and highly addictive drug. Opium harvest and production were possible for the British due to their successful conquest of large areas of India between about 1750 and 1800 (Spence 1990:130). The British quickly discovered that opium could be an effective cash crop in China for repairing perceived trade imbalances. Moreover, opium poppy grew abundantly in certain areas of India, and cheap labor to harvest the sap from incised pods and to boil it into a paste for smoking was readily available.

Opium flooded the Chinese black market, infiltrating every social class. Qing officials smoked opium to escape the boredom and pressure of government bureaucracy, women from wealthy families smoked to escape the realities of being trapped inside their homes with no means to obtain a formal education, merchants smoked while constructing business deals, students smoked for stress relief while taking the civil service examinations, soldiers on their way to war smoked to combat anxiety, and poor laborers licked opium tablets to ease heavy loads (Chang 2003:14; Spence 1990:130–131). By the early 1800s, enough Indian opium was entering Chinese ports to sustain over one million addicts (Figure 3.5), and the result was that a huge portion of the population was wasting away, often in a comatose-like stupor, slowing dying from addiction.

The Chinese government attempted to stop, or at least curb, the opium trade, but its attempts were largely unsuccessful. After 1800, for example, an edict was issued that outlawed the import and domestic production of opium (Spence 1990:131), and was quickly followed by the establishment of severe penalties for smoking the drug. Despite these efforts, by 1825 the flow of silver had reversed course and was leaving China to pay for opium shipments, creating a national financial crisis that resulted in a steadily higher tax burden on the poor (Spence 1990:149).

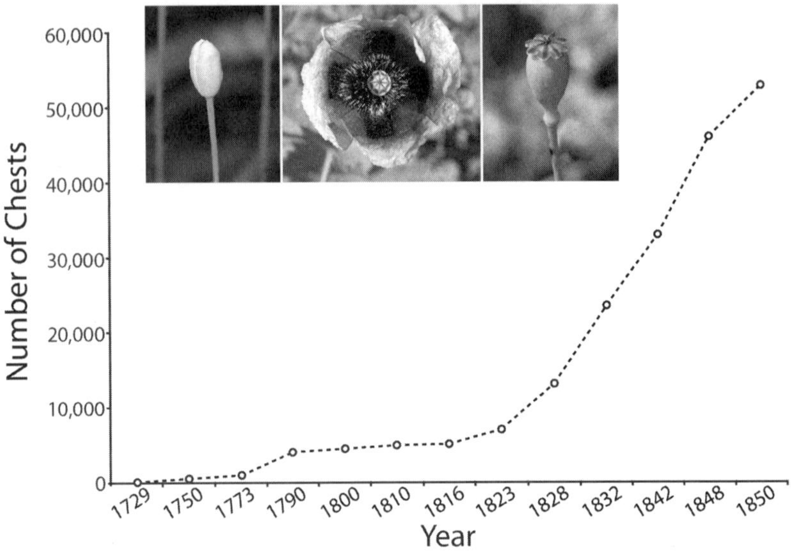

FIGURE 3.5. British sales of opium to China in the eighteenth and nineteenth centuries (based on Spence 1990:129 and Chang 2003:14). Each chest contained 130 to 160 pounds of opium. *Inset:* Composite photo depicting the three stages of an opium poppy (*Papaver somniferum*); *from left to right:* bud, flower, and fruit. (Open access images via http://commons.wikimedia.org)

By 1838, the emperor decided enough was enough and dispatched an imperial commissioner to Canton to end the opium trade once and for all. Chinese opium dealers were arrested, foreign residents were detained, and merchants were forced to surrender their opium stocks. The British government viewed these actions as the perfect excuse to break China's restrictive trade policies, under the pretense of protecting British residents in Canton. Its response was quick and decisive. A military fleet was dispatched to China and established a series of blockades at Chinese ports. Musket and cannon fire wreaked havoc on coastal cities and disrupted sea traffic and commerce throughout the country (Fairbank 1986:90–92; Spence 1990:155–156). Cities were captured and occupied, Chinese junks were sunk, and thousands of Chinese died during the conflict.

The end of hostilities, known as the First Opium War, came on August 29, 1842, when the Treaty of Nanjing was signed. Among the stipulations of the 12 main articles, China was forced to pay an indemnity of more than 20 million pounds to Britain, four coastal ports were opened to British trade, and Hong Kong was ceded to Queen Victoria (Hoe and Roebuck

FIGURE 3.6. Satirical cartoon published in 1840 showing an Englishman demanding that the Chinese emperor buy opium. Another Chinese man is dead on the floor, and English troops are in the background. The French text reads, "I tell you to immediately buy the gift here. We want you to poison yourself completely, because we need a lot of tea in order to digest our beefsteaks." (Open access image via http://commons.wikimedia.org)

1999:145–159). Although it was not explicitly addressed, the ultimate result was that China had little choice but to allow the continued importation of opium. The Chinese government shifted the burden of indemnities to the peasants (Figure 3.6), who paid exorbitant taxes and often were forced to sell all of their possessions to stay out of debt (Chang 2003:15–16).

The tipping point finally came in the nineteenth century when desperate Chinese peasants took up arms against the Qing government. The motivations for revolt were many, including oppressive governmental leadership, growing populations, environmental degradation, the outflow of the country's wealth (silver), mounting opium addiction, and foreign policy crises. Regardless of the triggers or mechanisms, the widespread rebellion left tens of millions of people dead, tore families apart, and ravaged communities. For example, the dissident leader Lin Qing, a working-class laborer from a small village outside the capital, recruited a cell of religious followers and persuaded them to rise against their Manchu oppressors in 1813 (Spence 1990:166–168). Lin's severed head was displayed in warning to future rebels after his devotees attempted to storm the Forbidden City and kill the emperor.

One of the most deadly rebellions began in poor, rural southeastern China in the 1840s. Its leader was Hong Xiuquan, a village schoolteacher who, after suffering the stress and humiliation of repeatedly failing his scholarly examinations, came to believe that he was the younger brother of Jesus Christ (Spence 1996). By 1847, Hong had attracted approximately 10,000 adherents, primarily from the lowest rung of Chinese society, based on an ideology of a new Christianity and the collapse of the Manchu hierarchy. Hong's revolution was a religious one, fought against demon Manchus who had turned their backs on the one, true god.

By 1850, tens of thousands of recruits were trained, armed, and loyal to Hong's cause. Late that year, they were tested in their first armed conflict and defeated Qing government forces that were attempting to suppress the movement (Spence 1996). Hong Xiuquan declared himself the Heavenly King of the Taiping Tianguo (Heavenly Kingdom of Great Peace) and marched north, seizing cities, recruiting new members, and building a vast supply of weapons, cash, and food (Spence 1990:172–173). The Taiping Rebellion, as it became known, was in full swing (Figure 3.7).

The conflict continued for 14 bloody years. At its most powerful, the Hong army established a capital at Nanjing and controlled a large portion of southern China, including 30 million people (Spence 1996). By 1864, however, imperial forces overwhelmed the rebels and retook Nanjing and the other major cities under Hong control. When all was said and done, the death toll of civilians and soldiers numbered in the tens of millions and the countryside was ravaged and depleted (Spence 1996). Poor farmers suffered the fallout, and many were forced to flee in search of jobs and arable land.

FIGURE 3.7. Artwork from 1886 depicting a scene from the Taiping Rebellion, the retaking of the city of Yuezhou from rebel forces by Qing troops. *Inset:* Rebel leader Hong Xiuquan. (Composite by author; all images open access via http://commons.wikimedia.org)

Escape to Gold Mountain

Nowhere was famine, starvation, and social unrest worse than in the province of Guangdong. In addition to these widespread plights, Guangdong was in the midst of a credit crisis, which began in 1847 when British banks cut off funding to warehouses along the Pearl River (Chang 2003:17). An outgrowth of the opium and foreign trade conflicts, the crisis left 100,000 laborers out of work and destitute for over a year.

Guangdong, located in southeastern China along the South China Sea, is an important trading center, connecting China with global commerce markets; Canton (Guangzhou) is its main port. Guangdong has also long supported a vibrant fishing industry along the coastline and the banks of the Pearl River. But during the nineteenth century, it became the launching point for poor laborers desperate enough to defy governmental law and escape civil war to shores abroad. The destitute escaped in droves to places such as Southeast Asia and Latin America. Many began new lives as rice farmers, fishermen, or merchants, but others were conscripted into lives no better or worse than what they had before. Between the 1840s and 1870s, for example, three-quarters of a million Chinese men were tricked or abducted by Chinese recruiters and sold into slavery. Labor shortages on colonial plantations in the Caribbean, South America, and Pacific Islands were filled by unwitting Chinese immigrants, an ugly period of Asian slavery that became known as the "coolie" trade (see Chang 2003:30–31; Hsu 2000:34; Pans 1994:67–69). The working conditions for coolies were horrendous, but the passage may have been even worse, with transit death rates estimated between 15 and 45 percent (Chang 2003:31).

In the middle of this chaos, one small discovery across the vast Pacific Ocean at a sleepy outpost, Sutter's Mill, forever changed the American West and the lives of generations of families worldwide. Pioneer James W. Marshall's discovery of several small gold flakes along the American River at the foothills of the Sierra Nevada on January 24, 1848, sparked the California Gold Rush (Rohrbough 1998:7).

Stories trickled into Canton of a land called *Gum Shan*, Gold Mountain, where untold wealth was readily available to anyone willing and able to make the dangerous journey across the Pacific. Impoverished Chinese farmers, fishermen, and laborers saw Gold Mountain as the solution to all their problems, and hope was ignited that, even with one small handful of gold, their lives and the lives of their families would be forever transformed. Scores of Chinese men begged, borrowed, and raised money to pay their passage to San Francisco. Unmarried men found wives and fathered children before they left, leaving these families behind but promising to send support and return home as soon as they made their fortunes (Chang 2003:19). They soon discovered, however, that the dream of Gold Mountain was very different from the reality.

CHAPTER 4

Chinese Immigrants in the Wild West

America in the mid-nineteenth century was a nation in transition. The Revolutionary War between Great Britain and the Thirteen Colonies had ended less than 70 years before, in 1783, and the women's suffrage movement was in full swing. The ethics and morality of slavery were being hotly debated and contested, and southern states were arguing for its expansion to newly established states and a reopening of the trans-Atlantic slave trade (Freehling 2007:10–20). The publication of the abolitionist Harriet Beecher Stowe's *Uncle Tom's Cabin* opened the eyes of the North, and the world, to the horrors of slavery. Add to this the fact that northeastern states had the largest cities (still relatively small compared with those of Europe and Asia at the time) and most of the factories and industrial development, while the southern states were dominated by low population densities and vast agricultural fields. What was left was a young country simmering with racial, political, and economic tensions on a rapid march to civil war.

The geopolitical map of California and America at the dawn of the California Gold Rush was also very different from what we know today (Figure 4.1). The Union consisted of about 30 states, and its western extent only reached Texas and several states bordering the Mississippi River. The Mexican-American War had recently been fought between 1846 and 1848 over the annexation of Texas, which Mexico considered part of its territory acquired after independence from Spain in 1821. The conflict spread into parts of New Mexico, Alta California, Baja California, and as far south as Mexico City; American forces captured Mexican strongholds and established crippling blockades (Crawford 1999). Mexico's military

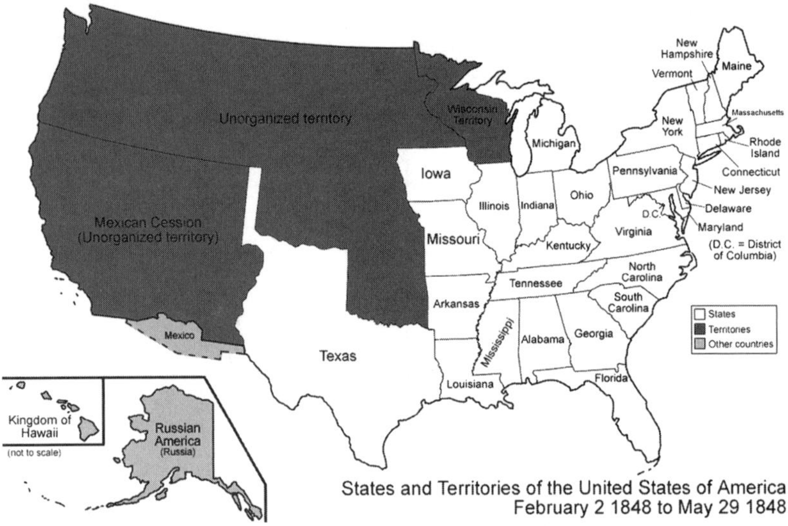

FIGURE 4.1. Map of the United States immediately following the discovery of gold in California. On February 2, 1848, Mexico ceded a large portion of its land to the United States as remittance for the Mexican-American War. (Open access image via http://commons.wikimedia.org)

found itself outnumbered and many of its largest cities occupied by U.S. forces. As part of the Treaty of Guadalupe Hidalgo, which ended the conflict, Mexico ceded the territories of Alta California and New Mexico in exchange for $15 million and the transfer of over $3 million of debt owed to U.S. citizens (Crawford 1999). California, then, was not officially a member of the Union but rather part of a large, unorganized (and mostly unregulated) territory.

This was the Wild West of Hollywood fame, a vast and largely unpopulated territory that brave and hardy pioneers settled beginning in the early nineteenth century. These pioneers were driven by the belief that they were destined to settle, civilize, and tame the wild continent as part of the "Manifest Destiny" ethos (Miller 2006). The same attitude steered the imperialist mindset of the American government, which threatened or declared war on its northern and southern neighbors in order to secure and expand its landholdings. Native Americans were seen as a threat to the divine Providence that drove westward expansion. Many tribes were forced to sign treaties, ceding the best of their lands and forced to relocate, often to isolated and desolate reservations. When the ambitions of Manifest Destiny met Native opposition, the most convenient solutions were frequently violence, disease, and murder. The "empty" West, then, was empty only because of the approximately 90 percent reduction of Native Americans, the result of Old World disease epidemics and Euro-American violence (see Denevan 1992; Ramenofsky 1987; Thornton 1987).

For colonizing-minded pioneers, the journey west was a dangerous and arduous one. An overland route required crossing multiple mountain ranges (the Rocky Mountains, the Sierra Nevada, the Wasatch, or others, depending on the route), vast plains, deserts, and high mountain passes. Following the Oregon Trail from Independence, Missouri, across the Continental Divide and then one of several routes to the coast, guided stagecoach or wagon trains usually took four to six months to reach their destinations (Rarick 2008:45–47). Other migrants opted to travel by ship and chose either the more-than-10,000-mile voyage around the southern tip of South America or a combined sea and land route down to Central America along the Atlantic, overland to the Pacific, and up the West Coast, once again by sea (Chang 2003:24–25). By any route, the risks were severe, and many never arrived.

Perhaps the most famous example of pioneer tragedy came during the summer of 1846 when the 87 members and 20 wagons of the Donner Party traveled an untested route between Fort Bridger, Wyoming, and the Humboldt River, Nevada (Grayson 1990:223). Before leaving their homes in Illinois, the Donner and Reed families (the founding members of the wagon train, with their numbers growing as the journey continued west) were persuaded to take a shortcut to California by a popular and influential guidebook, *The Emigrants' Guide to Oregon and California*, published by Lansford W. Hastings. Instead of heading north from Fort Bridger to Idaho's Fort Hall (the standard route followed by most emigrants), travelers were tempted by the Hastings Cutoff to save time and effort by heading south and west across the rugged Wasatch Range, through the Great Salt Desert, to the headwaters of the Humboldt River (Grayson 1990:224; Rarick 2008).

Unfortunately, the shortcut had never been traveled by wagon. Harsh terrain, heavy late October snow, and group infighting slowed their progress, and by early November, long after it had planned to arrive in the Sacramento Valley, the Donner Party found itself trapped by snow in the Sierra Nevada (Rarick 2008). Unable to continue until the snow retreated, the party made camp along what is now Donner Lake (Grayson 1990:228;

FIGURE 4.2. Historical newspaper photograph from 1866 near the Donner Party camp, established during the winter of 1846–1847. *Inset:* George Donner, who survived the Donner Party tragedy as a child. (Composite by author; both images open access via http://commons.wikimedia.org)

Figure 4.2). After several rescue attempts failed, cold and starvation quickly took their toll and most of the party had little choice but to resort to cannibalism to stay alive (Rarick 2008). By the time the final survivor was rescued, 40 members of the party had died, most from starvation (Grayson 1990). More than 150 years later, the Donner Party tragedy has become engrained in the folklore of early California and nineteenth-century westward expansion.

The oceanic journey for Chinese immigrants, dreaming of riches on Gold Mountain, was no less grueling. Most of the immigrants who could afford to pay for their passage across the Pacific worked with emigration brokers, called *k'o-t'ou* or *towkay* (Chang 2003:32). Brokers would arrange transportation to America and provide housing and protection in major port cities such as Hong Kong and Canton, dangerous places for the uninitiated. Brokers also provided a link between the immigrants and their families in China during their stay in America—arranging for mail, remittances, and news to cross the Pacific (Chang 2003:32).

For those who could not scrape enough money together to pay their passage, Chinese merchant middlemen organized a credit-ticket system, which became the most popular way to pay for transit to California. The immigrant would take on $40 to $60 in gold debt and pay a monthly interest between 4 and 8 percent (Campbell 1923:29; Chinn et al. 1969:14–15; Liu 1996). Firms in America collected the debts and arranged stateside employment (Campbell 1923:29), and immigrants were commonly burdened with this financial obligation for up to five years. Alternatively, Chinese immigrants entered into a contract labor agreement for a predetermined length of time. These contract laborers were originally described as "coolies," a word derived from the Chinese expression *k'u-li*, meaning "hard strength" (Chang 2003:30; Chinn et al. 1969:15). In nineteenth-century America, however, the word became a pejorative term to describe any Chinese laborer or, for many Euro-Americans, any man of Chinese descent.

Once the passage was paid and the terms established, the immigrants embarked on the difficult and dangerous trans-Pacific crossing to California. The voyage could take anywhere from about 30 days to three months, depending on weather conditions and whether the crossing was by junk, boat, or steamer (Chinn et al. 1969:16). During the early phases, immediately after the discovery of gold in California, much of the transit was aboard American- or British-owned ships that were able to reach San Francisco in four to eight weeks (Chang 2003:32). Chinese passengers spent most of the long trip below deck in the overcrowded steerage, crammed into makeshift wooden bunks that allowed precious little headroom (Kemble 1953:209, 214). Circumstances for the Chinese travelers varied, but disease and hunger were common. A young British journalist, Andrew Wilson, described the conditions for 400 Chinese immigrants aboard the American clipper ship *Bald Eagle* as it made a 41-day trip from Hong Kong to San Francisco: "[Chinese] passengers had just cause of complaint against the Contractor for their provisions, on account of the character of the beef.... Barrel after barrel of the beef was thrown overboard, as unfit for use.... The way in which the berths and cooking-places were put up in the *Bald Eagle* was disgraceful.... Many of them broke down, to the risk of heads and limbs; and the latter were so constructed that some days,

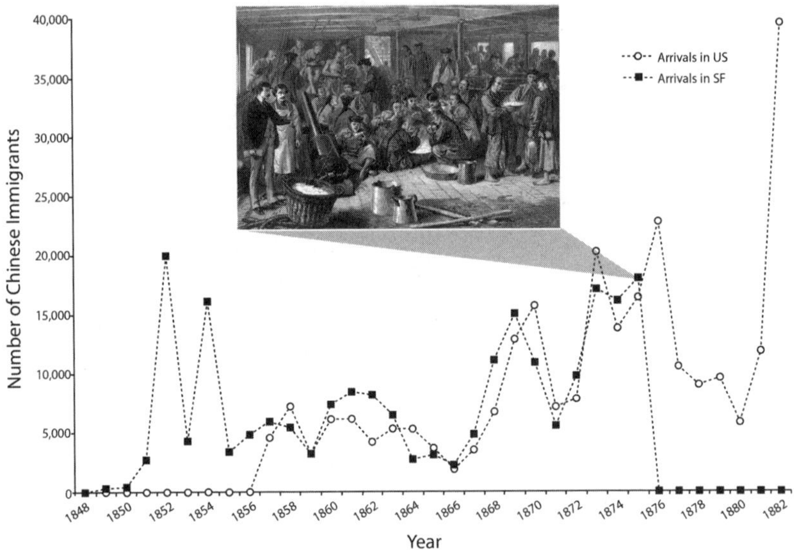

FIGURE 4.3. Number of recorded Chinese immigrants arriving in the United States and San Francisco from the beginning of the Gold Rush until the 1882 Exclusion Act (based on Coolidge 1909:498). *Inset:* Sketch of Chinese immigrants below deck of the steamship *Alaska* bound for San Francisco, ca. 1876. (Open access image via http://commons.wikimedia.org)

when the weather was bad, the emigrants got no food, as no cooking could be done, and they could not eat raw pork and unboiled rice" (Kemble 1953: 214). In some cases, these deplorable conditions ended in tragedy. Aboard the sailing ship *Libertab*, for example, abysmal health conditions resulted in the death of the captain and 20 percent of the 500 Chinese in transit from Hong Kong to San Francisco in 1854 (Chen 1981:23).

Braving and surviving the treacherous journey, Chinese immigrants came to San Francisco in search of Gold Mountain by the thousands (Figure 4.3). Before 1849, fewer than 50 Chinese lived in the continental United States, most of them scholars, merchants, sailors, or performers in urban port cities along the North Atlantic coast (Chang 2003:26, 103; Voss and Allen 2008:9). Only four years after the discovery of gold at Sutter's Mill, more than 20,000 Chinese immigrants arrived at the San Francisco customs house (Coolidge 1909:498). They dreamed of wealth and opportunity in America and a triumphant return to their families and homeland. But it was not long after arriving in San Francisco Bay that they confronted a sobering reality.

The City by the Bay at Chinese Arrival

Until the discovery of gold in California, historical San Francisco was not much more than a desolate area marked by sand dunes and rolling hills. In 1769, the area was "discovered" and explored by Junípero Serra, a Franciscan friar, and Gaspar de Portolá, a soldier, who led a Spanish expedition to establish missions and presidios up the Alta California coast to San Francisco Bay (Hittell 1878:41). For nearly 100 years, the area served as a military outpost and contained little more than a chapel, brush houses and huts, and a presidio (Chang 2003:34). In fact, San Francisco overtook Monterey as the coast's largest port only at the end of the Mexican-American War and, by the spring of 1847, barely contained 500 mostly Spanish-speaking residents (Paddison 1999:305, 309).

San Francisco was immediately and forever transformed by the discovery of gold. Tens of thousands of gold rushers descended on the city as the natural port for their journey to Sutter's Mill and mines in the Sierra Nevada. Veterans of the war with Mexico, survivors of Ireland's 1846–1847 potato famine, and European immigrants escaping revolutions in France and central Europe all sought opportunities for a new life, free from poverty and oppression. The historian Hubert Howe Bancroft estimated that 42,000 emigrants arrived along overland routes in 1849 alone, while arrivals by sea were assessed at 40,000 (Scott 1985:27). By the summer of 1849, the sleepy city by the bay was a Gold Rush town of 5,000 people, a town that would grow exponentially in the next three decades to rival the population, wealth, elegance, and architecture of the major East Coast metropolises of New York, Philadelphia, and Boston (Chen 1981:55).

Gold Rush arrivals to San Francisco Bay were confronted with a city struggling with unchecked population growth. Building supplies were in short supply, but hundreds of ships drifted abandoned in the harbor (Chen 1981:56). Their owners were impatient gold miners from all over the world. These ghost ships were hauled ashore, broken down, and used to construct shanties and stores, makeshift bridges and city streets.

The city's infrastructure grew by leaps and bounds as builders made a desperate attempt to keep pace with the influx of residents (Figure 4.4). San Francisco's first public square, Portsmouth Square, for example, was little more than a cow pen in 1848, surrounded by tents and adobe huts (Chen 1981:56). Less than 10 years later, the public space was bounded

FIGURE 4.4. An 1853 map of San Francisco, San Francisco Bay, and the surrounding areas by the United States Coast Survey. Note that the city extends for only about eight city blocks from the waterfront. Mission de San Francisco can be seen in the lower left corner. *Inset:* An 1849 handbill advertising the "quickest, safest, and cheapest" passage to California, aboard the steamship *Nicaragua* for $90 in 35 days. (Composite by author; both images open access via http://commons.wikimedia.org)

by stone and brick buildings, including hotels, offices, shops, restaurants, saloons, and gambling dens. The rest of the city followed a similar building trend: churches, government administration buildings, and courthouses were low priorities, and hotels, saloons, and casinos were in high demand. By 1853, San Francisco had 537 places that sold liquor, including 46 gambling halls, 114 taverns, and 48 brothels (Chang 2003:34; Eames 1995:48).

It was not only gold rushers and entrepreneurs who were attracted to San Francisco, but also outlaws, thieves, and crooks. The establishment of civic leadership, administration, and law enforcement lagged behind population growth, as was the case with many Wild West boomtowns—made famous by the Hollywood western. The initial lack of hotels, warehouses, and infrastructure forced many people to live in tents and leave their possessions exposed to thieves and the elements (Chen 1981:55). A court system and a police force were not instituted until 1850, and a California land office was not established until 1853 (Chang 2003:36). Without a government to regulate behavior and settle disputes, violence, robbery, murder, and mayhem were common, and conflicts were often resolved with fists, knives, and pistols. Between 1850 and 1856, 1,400 murders were recorded in San Francisco, not to mention other crimes such as robbery and assault (Chen 1981:55). At times, organized criminal gangs terrorized honest citizens, and vigilantes and mobs took to the streets to drive out violent criminals or hang them in public spaces (Chen 1981:55). These mobs cared little for due process or sufficient evidence, making them as dangerous as the criminals themselves.

The propensity for drinking, gambling, vice, and other seedy pastimes by the residents might be explained, at least partially, by the demographics of Gold Rush San Francisco. Most prospectors were young, either single men or married men who decided against relocating their families. This left the city largely devoid of females, and the men outnumbered the women 12 to 1 (Chang 2003:35). The demographics of San Francisco were very similar to census data from California at large. In 1850, 92 percent of California's 93,000 residents were male and 91 percent were between the ages of 15 and 40 (Milner et al. 1994:815). Gold Rush prospecting was a magnet for unattached male youths, predisposed to footloose, wasteful, and violent behavior. In 1850, gold rusher and Universalist minister Franklin Langworthy cogently described the leisure activities of a typical miner: "Packs of cards are kept in nearly every house, tent, and cabin, and are generally in use either for amusement or gambling. Checkers and chess are seldom played, these games requiring the player to exercise some little thought and reflection. Mental labor of every kind is altogether avoided, and looked upon with contempt by the great mass of Californians. Tippling and profane swearing are all but universal" (as cited in Dary 1995:203).

One of the most popular places for these men without women to spend their time and money was in brothels. In the earliest days of the Gold Rush, more prostitutes probably arrived in California than all the women in more respectable positions (i.e., housewives or servants) combined (see Gentry 1964:26). Almost all of California's pioneering prostitutes were of Mexican, Spanish, or French descent and arrived via Mexico or Central or South America (Gentry 1964:27). In the early days of the Gold Rush, fortunes of gold dust and nuggets were regularly found, and money for prostitutes was plentiful, making the world's oldest profession especially lucrative. Sex workers charged between $50 and $100 for their services (Jones 1998:152), between $1,550 and $3,100 in today's currency. Some enterprising prostitutes rode from mining camp to camp and made up to $50,000 (Jones 1998:152), over $1.5 million in today's currency.

After the initial months of easy surface gold, however, much of the wealth of Gold Rush California was made by capitalists and entrepreneurs who catered to the needs of the mines and miners. In San Francisco and throughout gold-country California, people could either work a mine or supply miners with essential goods and services, and sex was just one of these. Small business owners, many of whom began as prospectors, made fortunes selling basic products such as food, equipment, and clothing after colossal markups (Chang 2003:35) and offering transportation, banking, and remittance services (Chen 1981:56). In 1849, for example, a steam engine purchased for $2,000 on the East Coast sold for $15,000 in San Francisco (Holliday 1981:355). Women earned $50 to $60 a day ($1,550 to $1,900 today) washing clothes or $30 a day ($900 today) cooking (Holliday 1981:355). One woman reported $600 a day ($18,000 today) in profits from baking pies, most over an open campfire (Jones 1998:153).

The legacy of gold-mining entrepreneurship can still be seen today. Levi Strauss, for example, identified the need for tough, durable work clothes for miners. The pants he created out of tent canvas reinforced with copper rivets at the pockets and other points of stress famously became the basis of a global fashion empire (Downey 2007). John Studebaker built his automobile fortune from a wheelbarrow manufacturing business (Chen 1981:56). Many of the later railroad barons, such as Leland Stanford, Collis P. Huntington, and Mark Hopkins, began accumulating their fortunes as California Gold Rush merchants.

San Francisco and the Gold Rush drew not only miners and merchants, thugs and thieves, but, paradoxically, intellectuals. The city rapidly became the most cultured on the West Coast and boasted more college graduates than any other in the country (Chang 2003:36). A dozen newspapers were up and running by 1853, and theater performances were a growing attraction. At the same time that people yearned for gold and wealth, they also sought new ideas and cultural experiences. The City by the Bay was rapidly becoming a cosmopolitan mix of rich and poor, educated and uneducated, aristocratic and blue-collar, which gave birth to a vibrant and complicated cultural landscape.

The early Gold Rush years in San Francisco and California were largely a halcyon time. Although an undercurrent of violence and corruption permeated the boomtown, there was enough wealth and opportunity to keep the majority of folks satisfied. San Francisco was also a global city: by 1853, more than half of its population was foreign-born (Jeffrey 1998:143), a melting pot of the American West. Most people were too busy to notice the sudden and growing arrival of Chinese immigrants. They were simply another group of weary travelers and prospectors who came for much the same reasons as everyone else. Gold Rush fever had cast its spell, and few people could see through its golden haze. But as soon as gold became more difficult to find and competition increased, the air of openness and acceptance evaporated and Californians looked for someone to blame for their unrealized dreams of riches. The Chinese were easy scapegoats.

Gold Mountain, Railroads, and Opportunities at the Margins

Chinese prospectors began to arrive in San Francisco in relatively low numbers in the early years of the California Gold Rush (see Figure 4.3). Fewer than 1,000 reached Gold Mountain before 1851. Many San Franciscans regarded these early émigrés with interest, curiosity, and admiration. One resident, for example, commented on the "quietness and order, cheerfulness and temperance" of Chinese immigrants, glowingly describing their work ethic and "cleanliness" (as cited in Chinn et al. 1964:9). In 1850, a reception was held in downtown San Francisco for the "China boys," celebrating their contributions to the California labor market (Siu 1987:44–45).

The relatively small numbers of Chinese arrivals made them more of a novelty than a threat at first. The earliest arrivals probably also included more merchants and businessmen than laborers and gold prospectors. An 1884 *Overland Monthly* article recounts one resident's perception of the first Chinese in San Francisco: "It is enough to know that in the fall of 1849 the Chinese in San Francisco numbered several hundred. They were not laborers who came; not of the coolie class, at least. Very few of them went to the mining district, and the writer never saw, never heard of a Chinaman digging for gold in any of the placers that year" (O'Meara 1884:477). This account is somewhat exaggerated since many Chinese immigrants signed labor contracts as cooks and tailors to pay for their passage to California.

Between 1850 and 1852, the number of Chinese arrivals rose precipitately, reaching over 20,000 in 1852. Most of them quickly set out to find their fortunes in Gold Mountain, and by the early 1850s, approximately 85 percent of all Chinese immigrants in California were involved in mining (Chang 2003:38). Later waves of immigrants had the advantage of Chinese networks, whose members waited to greet them at the docks and facilitated their transition into American society. Following their collectivist and Confucius ethos, prosperous Chinese merchants organized social and welfare organizations to find shelter, jobs, and interpreters for newcomers. Clan (family) associations formed to help immigrants of one surname, and were quickly followed by district associations that united immigrants originating from one district or county who spoke the same dialect of Chinese (Chen 1981:27). Eventually, these associations merged into the Six Companies (Figure 4.5), or the Chinese Consolidated Benevolent Association (CCBA), and quickly became a powerful political organization in San Francisco and California, advocating for Chinese rights and providing lending, health, education, and funeral services. A young nineteenth-century Chinese immigrant, Fatt Hing, described his first contact with the Six Companies as he took his first steps onto American soil and the docks of San Francisco, recounting the words of the head of the Six Companies:

I am Wong Wing Dock, chairman of the Six Companies. We came as you came on board one of those ships, and we came for the same purpose—to seek gold. When we set foot on these shores, however, there were no Chinese faces to greet us, for we were among the first to arrive. Weak and wearied from our long journey, we were bewildered and lost. We did not know where to turn for shelter or food. Fortunately, there were enough of us so that some set to putting up these houses while others looked for food. We followed the white men into the hills and found out how they sifted the sands for gold and we did the same.

One valuable lesson we have learned and which you will soon appreciate is that we must stick together and help one another, even though we are not kin. That is why we have formed this organization called the Six Companies representing the six districts which most of us come from. Our compatriots have honored me by choosing me chairman.

You will always find food and shelter here among us. Any needed supplies can be purchased from the shops on this street. No doubt you will want to send a letter home to let your families know of your safe arrival. Elder brother Leong is a learned man, and he will help you write your letters.

When you have earned money from your diggings or from your wages, you will pay dues into the company fund. This fund helps us to maintain the company headquarters and helps us set up an orderly system to take care of our own.

We are Chinese in a land of foreigners. Their ways are different from our ways. Their language is different from our language. Most of them are loud and rough. We are accustomed to an orderly society, but it seems as if they are not bound by any rules of conduct. It is best, if possible, to avoid any contact with them.

Try not to provoke the foreigners. But you will find that they like to provoke us. We are comfortable in our loose cotton jackets and trousers and we are used to going barefooted. They like to wear rough coarse clothing with high-laced boots. They cut their hair short and let the hair grow on their faces. We wear our hair long and braided and we shave the hair from our faces. Since we all want to return to our homeland, we cannot cut off our queues. (as cited in Sung 1967:24–25)

FIGURE 4.5. Historical photograph of officers of the Chinese Six Companies in San Francisco. (Open access image via The Bancroft Library, University of California, Berkeley, call number BANC PIC 1905.17500 v.29:104—ALB)

Until the end of the nineteenth century, the CCBA represented immigrants from two dozen districts in Guangdong Province, where the vast majority of newly arriving Chinese originated (Chin 1981:27). These organizations provided basic necessities and helped Chinese immigrants acclimatize to their new surroundings. The simple act of providing boots, shovels, and picks was invaluable for facilitating and expediting their journey to Gold Mountain.

By 1851, when the Chinese began arriving by the thousands, San Francisco was one of the largest cities in the United States, with more than 30,000 residents (Chang 2003:34). Competition for mines and gold was fierce, but the Chinese, like many others, made their way to the mountains and foothills to carve out their niches. For most, this meant working from sunrise to sunset, bent over a pan, tediously separating gold dust and nuggets from gravel and dirt. Using a pan or other simple device to collect gold particles that have eroded into streams, hillside colluvial deposits, and glacial till is known as placer mining (Chung 2011:10; Jones 1998:133–134). More complicated methods, such as rocker and sluice boxes, which required several workers to operate, were also regularly in use.

The Chinese soon earned a reputation for hard work, tenacity, and cooperation, gaining the respect and attention of many scholars and travelers. The wealthy Scotsman John David Borthwick, for example, traveled through California between 1851 and 1854 and described life in San Francisco and the gold mines. He remarked that Chinese miners often worked in rugged, isolated areas where they could avoid conflict with Euro-American miners (Chung 2011:11–12). Their camps generally were well organized, scrupulously clean, cooperative ventures, which allowed for amazing productivity and, many times, successes (Tong 1994:13). Even Mark Twain commented on these traits: "They are quiet, peaceable, tractable, free from drunkenness. A disorderly Chinaman is rare, and a lazy one does not exist" (as quoted in Chang 2003:39). At the same time, however, many competing Euro-American miners viewed their Chinese counterparts with racist contempt. Borthwick (1857:144–143) illustrates:

> [The Chinese] are an industrious set of people, no doubt, but are certainly not calculated for gold-digging. They do not work with the same force or vigour as American or European miners, but handle their tools like so many women, as if they were afraid of hurting themselves. The Americans called it "scratching," which was a very expressive term for their style of digging. They did not venture to assert equal rights so far as to take up any claim which other miners would think it worth while to work; but in such places as yielded them a dollar or two a-day they were allowed to scratch away unmolested. Had they happened to strike a rich lead, they would have been driven off their claim immediately. They were very averse to working in water, and for four or five hours in the heat of the day they assembled under the shade of a tree, where they sat fanning themselves, drinking tea, and saying "too muchee hot."

Chinese miners worked tirelessly, chasing riches and dreaming of a return to China as wealthy conquerors. Hard work and luck helped some of these dreams come true (Chang 2003:40–43), but the Chinese were also resourceful and creative. For example, Chinese miners used a traditional shoulder pole to carry heavy equipment (Figure 4.6) and modified their sluice boxes so they could easily move from place to place in search of gold (Chung 2011:10–11). Archaeological research has produced instances of ingenious improvisation, such as a sluice box from Idaho repaired with metal strips from opium tins (Liestman 1998:20). The Chinese learned to adapt quickly and efficiently, to work in areas that Euro-Americans

FIGURE 4.6. Images of nineteenth-century immigrant Chinese gold miners: (*A*) a Chinese man carrying a sluice box and mining equipment, ca. 1860 (Open access image via http://commons.wikimedia.org); (*B*) a Chinese placer miner in the American West (Open access image via the Bancroft Library, University of California, Berkeley, call number BANC PIC 1996.001:88—ALB); (*C*) a group of Chinese miners working together at a mine in California (Open access image via the Bancroft Library, University of California, Berkeley, call number BANC PIC 1905.17500 v.29:129—ALB). (Composite by author)

believed inaccessible or unproductive, and to scavenge at abandoned claims. Their hard work, frugality, and ingenuity allowed them to capture millions of dollars of gold from California. Some returned to China in short order, and others found success by investing their fortunes in farms and businesses in the American West (Chang 2003:45). But the dreams of riches were realized for only a select few; most Chinese miners, like their Euro-American counterparts, toiled in the mountains, rivers, and foothills of California for small returns. Their dreams of wealth and prosperity were as elusive as the tiny gold nuggets and flakes they pursued in the rugged, unforgiving wilderness of the Sierra Nevada.

Problems for the Chinese began to mount as competition grew for an increasingly smaller amount of gold. More and more miners swarmed into the gold fields, and by 1852, an estimated 100,000 people were actively engaged in the industry (Jones 1998:135). New mechanical techniques were adopted to divert rivers, construct dams, and move massive amounts

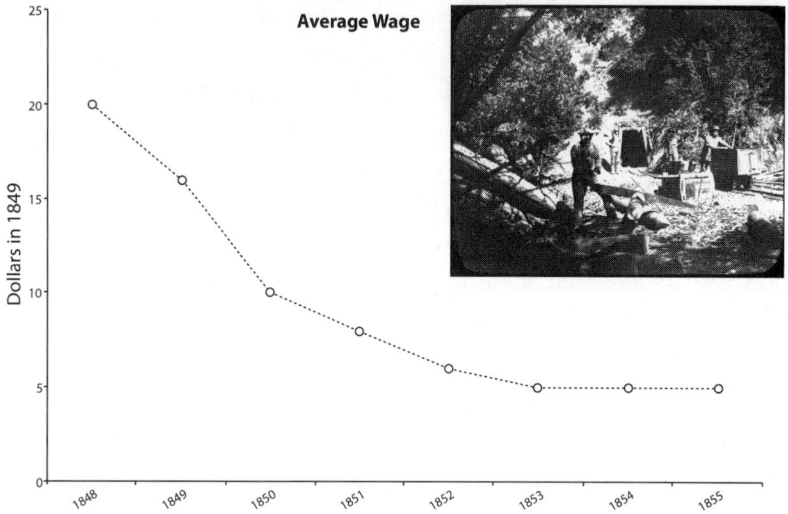

FIGURE 4.7. Average daily wages for white miners during the height of the California Gold Rush (based on Jones 1998:139). *Inset:* Four men outside a California gold mine in the Sierra Nevada during the late nineteenth century. (Open access image via http://commons.wikimedia.org)

of dirt, requiring a considerable investment of labor and capital and making placer pan and sluice box mining obsolete. Even with these new techniques, steep declines in the extraction of gold were felt by 1853 (Rice et al. 1988:165–185). The Chinese were pushed to the margins of mining society, left to pick already-worked or peripheral areas that yielded small, if any, returns (Jones 1998:139).

As returns began to diminish, Euro-Americans voiced their growing displeasure with toiling in the gold mines day after day and month after month, hiking up and down steep mountainsides, excavating boulders from rivers, and endlessly swinging picks and shovels—all without realizing the bonanza they believed they deserved. Average wages for white miners began a steady decline shortly after the start of the Gold Rush, as more and more laborers rushed in to make their fortunes (Figure 4.7). Although Chinese miners dealt with the same harsh realities of unrealized dreams and their daily wages were even lower, Euro-American miners blamed foreigners and Johnny-come-latelies, arguing that they were carrying off American wealth to foreign lands. Lawmakers agreed, and the California legislature passed a foreign miners' tax of $4 a month in 1853 (Tung 1974:8–9). The law was intended to include all foreigners, but in reality it became a tax on immigrant Chinese miners, even though Irish immigrants outnumbered the Chinese, and Germans were a close third (Sung 1967:27). Given their unfamiliar hairstyles, clothing, customs, languages, and physical ethnic characteristics, Chinese "foreigners" were much easier to recognize than ethnic groups of European ancestry, and the Chinese became easy scapegoats. Although the law was later repealed as unconstitutional, the Chinese paid 85 percent of the collected taxes and new Chinese immigrants to California were dramatically reduced that year since they could not afford the financial burden (Sung 1967:27–28; Tung 1974:9).

The racist sentiment against Chinese immigrants is not surprising when put into historical perspective. California joined the Union as a free state in 1850, but slavery was still alive and well throughout the southern states. The Americans who flooded California during the Gold Rush came from every corner of the country and brought with them their racial antipathies. Many southerners wanted to maintain the tradition of slavery while many northerners did not, but both were equally unwelcoming to blacks (Chen 1981:44–45). As part of the state constitution, for example, blacks could not vote, hold public office, settle on a homestead, serve on a jury, attend schools with whites, or testify in court (Chen 1981:45). Chinese miners and laborers, "coolies," were the equivalent of black southern slaves, perhaps not by law but in the court of public opinion, and suffered many of the same injustices as free blacks in California (Sandmeyer 1991:26).

In 1849, a law was passed that no person of color could provide evidence for or against a white person in court (Chen 1981:45). Because Chinese immigrants were considered persons of color, they could do little to counter the crimes and injustices perpetrated against them by whites. Stories of white miners burning Chinese tents and mining equipment became commonplace. Chinese camps were regularly raided by whites, who stole their gold stashes and murdered or tortured the occupants. Elmer Clarence Sandmeyer (1991:45) described the fallout from the new legislation: "It is obvious that [the passing of this law] opened the way for almost every sort of discrimination against the Chinese. Assault, robbery, and murder, to say nothing of lesser crimes, could be perpetrated against them with impunity, so long as no white person was available to witness in their behalf."

FIGURE 4.8. Various occupations of Chinese immigrants in San Francisco during the nineteenth century: (*A*) a Chinese restaurant on Dupont Street, between 1868 and 1900 (Open access image via http://commons.wikimedia.org); (*B*) image published in *Harper's Magazine* in 1882 of the Chinese Exchange (Open access image via the Bancroft Library, University of California, Berkeley, call number MTP/HW: Vol. 26: 173); (*C*) image published in *Illustrated San Francisco News* in 1869 of the Chinese Cigar Manufactory on Merchant Street (Open access image via the Bancroft Library, University of California, Berkeley, call number fxF850.I42 Vol. 2:1, 21). (Composite by author)

These difficulties drove Chinese miners to the margins of the gold industry and, for many, to alternative economic opportunities (Figure 4.8). Some Chinese miners, for example, returned to San Francisco and opened restaurants catering to the appetites of the growing immigrant Chinese population. Chinese restaurateurs identified themselves with triangular flags fashioned from yellow silk, and many earned a reputation for excellent cuisine among not only Chinese immigrants but also Euro-Americans (Chen 1981:57–58). Gold miner William Shaw (1851:42) wrote, "The best eating-houses in Francisco are those kept by Celestials, and conducted Chinese fashion; the dishes are mostly curries, hashes, and fricasees, served up in small dishes, and . . . exceeding palatable." To be fair, others were less glowing in their praise of Chinese restaurants and cuisine. While traveling around the world, Reverend Otis Gibson (1877:70–71), for example, described the Chinese restaurants in downtown San Francisco on Dupont and Jackson Streets: "The Chinese cooking is more like the French than the English. They are fond of cutting every thing up fine, and mixing different things together. Their meats are usually well cooked, and some of their pastry is light, though not generally so. The principal drawbacks to the enjoyment of a real Chinese dinner are two: the inability of Americans to use chopsticks, and the fact that many of the dishes taste of rancid oil or strong butter." Gibson (1877:71–72) went on to describe, however, his astonishment at the voracious appetite of his American dinner guest at one of these Chinese restaurants. In the middle of the meal, his fellow patron dispensed of his chopsticks and began scooping the food into his mouth with his fingers, lamenting that, if not for a lecture he had to present, he would continue eating.

Chinese immigrants also found financial success in California by opening laundries that catered to white miners. In the mid- to late nineteenth century, washing and ironing clothing was tedious, laborious, and generally considered women's work. Boomtown San Francisco had precious few women and even fewer laundry services. The best option for many Californians, Chinese or white, was to send their laundry via steamship to Hong Kong or Honolulu for cleaning and ironing (Chang 2003:49; Chen 1981:58). As might be imagined, the prices for this service were exorbitant, ranging from $8 to $12 (between about $215 and $330 today) for a dozen shirts that would take up to four months to make the round trip across the Pacific and back (Chen 1981:58). For many laborers and gold miners, the price of laundry was equal to a day's wages or more.

Failed gold miner Wah Lee opened the first Chinese laundry in California in 1851 at the corner of Grant Avenue and Washington Street in downtown San Francisco (Siu 1987:46). For $5, Wah Lee would wash and iron a dozen shirts in a fraction of the time it took to send them overseas (Chen 1981:58). Within 20 years, San Francisco supported 2,000 Chinese laundries, and by 1880, an estimated 7,500 Chinese immigrants were involved in the industry (Chen 1981:58).

Enterprising Chinese immigrants successfully expanded into a variety of other businesses. Chinese merchants opened shops and distributed goods such as fresh flowers, vegetables, and fruits door to door (Chen 1981: 58–59). Chinese curio stores sold a variety of collectibles such as porcelain vases, ivory and jade artwork, chess pieces, fans, kettles, shawls, scroll paintings, teapots, and other Asian products, which became highly sought after by Victorian households and white miners (Borthwick 1857:75; Chang 2003:49; Chen 1981:59), as well as other Chinese immigrants.

Perhaps most famously, immigrant Chinese laborers abandoned California's gold mines for steady employment (at least temporarily) building the transcontinental railroad. Until the completion of this vital transportation system, there was no safe and reliable route linking the east and west coasts. Mail, products, and people would take months to reach California, if they ever arrived at all. Two overland routes existed, one over the treacherous Rocky Mountains and the other, southern route through the heart of hostile Indian (Apache and Comanche) country (Chang 2003:53–54). The sea route, although safer, took even longer aboard boats launched from the Gulf of Mexico or the Atlantic Ocean which had to round the southern tip of South America and travel up the Pacific Coast.

Even though the country was a year deep into the bloody fighting of the Civil War, Congress, with President Lincoln's support, passed legislation in 1862 to fund the completion of the transcontinental railroad from its western extent in Omaha, Nebraska (Chang 2003:54). Chinese immigrants provided the labor backbone for this massive feat, one that many engineers believed was impossible or far too dangerous to even attempt (see Ambrose 2001; Bain 1999). Thousands of Chinese toiled on the rail lines, and dozens died in the effort. A variety of white and other immigrant laborers were necessary for completing the railroad in only seven years, but Chinese sweat, blood, ingenuity, and fidelity were essential for the speed and success of the project. Ironically, these contributions and successes added to the heightening of the anti-Chinese movement in California and around the country (more in Chapter 7). All this work was done at daily wages below those of their white counterparts, saving the railroad companies and the nation millions of dollars (in 1865 rates) and untold time measured in days, months, or years.

Chinese Fisheries

One of the most successful, and perhaps less well known, ventures of Chinese immigrants in the mid-nineteenth century was the founding of California's saltwater fishing industry. While their origins are hazy, Chinese fisheries in California probably came about in several different ways. Many historians have long believed that the earliest Chinese fishermen were gold miners who founded the industry after the Foreign Miners Tax of 1854 was instituted, making gold panning financially untenable (e.g., Chen

1981). Equally likely, however, is that the earliest Chinese fishermen came to California with the intention of building a fishing business or that they never pursued mining at all after seeing the untapped coastal resources of San Francisco Bay (Bentz and Schwemmer 2002:151; Lydon 1985:36). Since the vast majority of nineteenth-century Chinese immigrants came from Guangdong Province, a coastal area of southeastern China whose economic foundations were fishing, shipbuilding, and trade, it is likely that many were experienced or expert fishermen and came with the intention of fishing. Even if the earliest Chinese fishermen did not emigrate to the United States with this aim, many arrived with the skills and knowledge necessary to initiate the industry. Census records reveal, for example, that the residents of the Chinese shrimp fishing village at Point San Pedro on San Pablo Bay, just north of San Francisco, originated from Guangdong Province, where shrimp fishing was common (Nash 1973:186; Figure 4.9).

Regardless, the first Chinese fishing operations were launched sometime between 1849 and 1853, and the earliest documented fishing camp was established at Monterey in 1853 (see Armentrout-Ma 1981:142; Bentz and Schwemmer 2002:140; Chen 1981:97). Chinese fishermen discovered a California coastline overflowing with underexploited marine resources, many of which were considered delicacies in China. The waters of San Francisco Bay, the Sacramento and San Joaquin Rivers, the Monterey area, and San Diego Bay became commercially important fishing centers, and small Chinese settlements rapidly grew into permanent fishing villages. By the late 1880s, there were at least 20 fishing villages at San Francisco and San Pablo Bays, "one at Tomales Bay, four in the Monterey-Pacific Grove area, two near San Diego, one on the island of San Clemente and others on the Sacramento River" (Chen 1981:98). All told, more than 2,000 fishermen made their living along the California coast.

Shrimp, shark, redfish, rockfish, sturgeon, smelt, sole, barracuda, squid, clams, crabs, lobsters, and many other commercially important species were abundantly available and sold fresh (often live) to local markets and shipped to Chinese communities up and down the West Coast. Dried fish were used to provision Chinese gold miners in the interior (Chen 1981:98). The bulk of these commercial marine products, however, were dried and exported to mainland China, often to replenish overexploited fisheries in the immigrants' homelands, along with other ports around the Pacific Rim. Chinese merchants in Chinatown usually handled these

FIGURE 4.9. Chinese shrimp fishing village at Point San Pedro in San Francisco Bay in 1889. *Inset:* Chinese fishermen mending nets on one of the large drying areas on the slope behind the structures. (Composite by author; both images open access via http://commons.wikimedia.org)

transactions and arranged the sale and shipment of goods (Bentz and Schwemmer 2002:141). This commercial market for fish, which continues to be an important industry in California today, was pioneered by Chinese merchants and introduced many new marine products to Euro-American palates.

Many of the same traits that contributed to Chinese successes in other American endeavors also shaped their rapid dominance and expansion of the fishing industry. Chinese fishermen were hardworking, opportunistic, and resourceful. They employed efficient, traditional technologies adapted for use in the California fisheries. Technologies such as bag nets, troll lines, and traps were often innovative and exceptionally effective (Bentz and Schwemmer 2002:144). Bag nets used in the San Francisco Bay shrimp fishery, for example, were imported technologies from China (Nash 1973). Costing $12 apiece, these nets helped the Chinese capture the lion's share of the five million pounds of shrimp sold in San Francisco in 1871 (Chen 1981:99). In the waters off Monterey, Chinese fishermen worked in teams to trap squid during the dead of night. One boat would suspend a torch in a wire basket over the water with a pole, while two other boats encircled the shoals of squid attracted to the light with a giant purse seine net (Chen 1981:103).

Perhaps the most important traditional technology for Chinese success in commercial fishing was watercraft. Immigrant Chinese shipwrights constructed junks and sampans for fishing and transporting cargo throughout California. These vessels were built from native California redwood but with traditional Chinese construction techniques. The diversity in junk and sampan construction suggests that the vessels were adapted to meet the specific needs of the owners and the challenges of the California fishery rather than being built from rigid plans (see Van Tilburg 2007).

The term "junk" refers to a diverse range of vessels, but they were typically large, seaworthy crafts that had one or more sails and a rudder. They were effective craft for traveling to offshore fishing grounds and transporting cargo up and down the coast to market. Sampans were smaller, relatively flat-bottomed vessels built without decks and propelled by a small sail, oars, or poles (Bentz and Schwemmer 2002:147). Sampans were effective for local fishing and the transportation of smaller loads to market, but less suitable for open-water voyages.

One of the very successful commercial fisheries developed by nineteenth-century immigrant Chinese fishermen was for intertidal black abalone. Although abalone was an important source of food and tools for Native Americans for millennia, after European arrival the fishery went unexploited, at least initially, since Euro-Americans did not recognize its commercial potential or find abalone meat appetizing. In the next two chapters, I explore why this fishery was left unexploited (until the Chinese), what we know about the history of Chinese abalone fishing in California, and how the archaeological record on the Channel Islands provides documentation of the fishery. These chapters set the stage for a deeper understanding of abalone exploitation in California. The Channel Islands archeological record may hold the key to understanding how we can better manage this heavily overfished and rapidly disappearing marine resource.

CHAPTER 5

The Archaeology of Chinese Abalone Fishing in Southern California

The genesis of commercial abalone fishing in southern California was possible only because of the timing of a number of historical events, which created the perfect storm for abundant abalone stocks. The first seminal event coincided with the death of the famous Danish explorer and Russian Navy officer Vitus Jonassen Bering. Bering is best known for his explorations of the northeast Asian and western North American coastlines. The Bering Strait, the Bering Sea, Bering Island, Bering Glacier, and the Bering Land Bridge were all posthumously named in his honor.

In 1741, Bering and his crew became shipwrecked and stranded on the isolated and uninhabited Commander Islands, the western-most extension of the Aleutian Islands, just east of the Kamchatka Peninsula in the Russian Far East (Frost 2003; Figure 5.1). Starving and desperate, several crew members trudged down the beach and clubbed four sea otters, making a welcome meal of nearly every part of the animals (Ellis 2003:138–139). Unfortunately, Bering was too sick to eat and died a month after their ordeal began.

Starvation and scurvy claimed nearly half the crew, but the remaining 46 members salvaged wreckage from their ship, constructed a small boat, and successfully sailed back to Kamchatka. They crossed 60 miles of open, frigid ocean in 13 days, aboard a leaky vessel that had to be bailed around the clock (see Ford 1966). The crew jettisoned nearly all their possessions and equipment to keep afloat, but one crew member, Fleet Master Khitrov, managed to squirrel away a collection of prime sea otter pelts. After arriving ashore, Khitrov sold the contraband at exorbitant prices to Chinese merchants.

FIGURE 5.1. A nineteenth-century illustration by an unknown artist of the shipwreck of Vitus Jonassen Bering and his crew on the Aleutian Islands in 1741. *Inset:* Portrait by an unknown artist in the early to mid-eighteenth century of Vitus Jonassen Bering. (Composite by author; both images open access via http://commons.wikimedia.org)

There is some debate over the authenticity of this sea otter pelt story, which may have been fabricated (or exaggerated) by later historians to explain the return of Russian fur hunters to the Commander Islands (see Ellis 2003:140), but the point remains. Shortly after the return of Bering's surviving crew, Russian sealers and hunters began the wholesale slaughter of otters and other sea mammals in the North Pacific. This nearly marked the end of otters, and it was largely luck and environmental circumstance that saved them (and many other sea mammals) from extinction.

The mass slaughter of sea otters was driven largely by the market for their pelts in China (Ogden 1941). More than any other animal fur, otter pelts were cherished by wealthy Chinese for their warmth and beauty, and the fur became an important sign of status in Chinese society, adopted by the imperial family, aristocrats, and other wealthy classes. Women wore otter capes, fur belts and sashes adorned with strings of pearls, and silk gowns bordered with otter fur (Ellis 2003:140). For over 100 years, otter fur dominated North Pacific commerce and trade. By 1790, a single otter pelt could fetch between US$80 and US$120 in China (Ellis 2003:140). As

FIGURE 5.2. Graphs depicting the relationship between the depletion of sea otters in the North Pacific and the price of sea otter skins at Canton between 1805 and 1833. *Right y-axis and open circles:* Price in Spanish dollars of high-quality sea otter skins at Canton (see Gibson 1992:317). *Left y-axis and closed circles:* Sea otter fur imports by American vessels at Canton (see Gibson 1992:315). *Inset:* A California sea otter. (Open access image via http://commons.wikimedia.org)

the animals were hunted toward extinction and became more rare and harder to capture, prices rose and pelts became even more valuable and sought after by hunters (Figure 5.2). Even North American beaver pelts, the heart of the continental fur trade, could not compete with the elegance, opulence, and status of otter pelts. By 1829, for example, a single sea otter skin was worth 10 beaver pelts (Gibson 1992).

Exorbitant prices created a frenzy akin to the California Gold Rush (Ellis 2003:142–143). Untrained Russian fortune seekers flooded the North Pacific, and the successes of the earliest hunters fueled the ambitions of many more. The hunt quickly spread east across the Aleutian Islands and south down the coast. Otter furs were the prized possession, but fur seals, foxes, and other animals were also targeted (Ogden 1941). By 1812, the Russians established their southernmost sea otter hunting and trading outpost, Fort Ross, 75 miles north of San Francisco at Bodega Bay (Lightfoot 2005; Lightfoot et al. 1991). With the help of Aleut hunters brought from the north, California's sea otter population was quickly decimated. By the eighteenth century, Spanish and American hunters were following suit in Baja, Alta, and central California. They moved from south to north, competing with the Russians and Aleuts (and each other) in the frantic race for the valuable pelts. Their sea otter furs, along with the pelts hunted by the Russians, were bound for the Chinese markets in Canton, often via Hawaii. By the time the Monroe Doctrine, a policy that outlawed the colonization of American land by European interests, was introduced in 1823, adult sea otters were a thing of memory and only four yearlings could be found (Ellis 2003:144).

Approximating the number of sea otters before the commercial harvest and during the eighteenth and nineteenth centuries is exceptionally difficult (see Ogden 1941:140), but estimates suggest that between 500,000 and 900,000 otters were killed between 1742 and 1910 (Bodkin 2000:83). The combined blitzkrieg coming from both the north and south was devastatingly effective, and by 1910, when the U.S. government instituted federal protections, sea otters in the North Pacific were in serious trouble. Fortunately, their numbers in both the Far North and California have rebounded, at least to some degree.

In southern California, otters have been functionally extinct since the early 1800s, meaning they no longer play a significant role in the ecosystem, and they are listed as threatened under the U.S. Endangered Species Act. California sea otters were completely eradicated from Channel Island waters by the mid-1800s, surviving only as a small relict population along the rugged and remote Big Sur coast (Ogden 1941:7–8), an area where fur hunters never completely eradicated them because of its inaccessibility.

Otters are voracious predators of shellfish, including sea urchins, abalone, and mussels. Because they lack the thick, insulating blubber of other sea mammals, otters consume 25 percent of their body weight in shellfish meat daily. One of the outcomes of the sea otter fur trade, then, was the release of adult abalone (and other shellfish) from one of its two primary predators. For decades, abalone flourished as a direct result of otter population declines.

The other major predator of adult abalone in southern California, the relatively dense populations of Chumash and other Native American groups that occupied the mainland and islands for millennia before European contact, was decimated by diseases introduced by early Spanish explorers. A variety of circumstantial evidence suggests that Old World disease epidemics (e.g., smallpox, chicken pox, measles, mumps, influenza, bubonic plague, typhus, yellow fever) and indigenous hemorrhagic fevers had significant impacts on Native American populations in California as early as the 1500s (Erlandson and Bartoy 1995; Preston 1996, 2002). Archaeological evidence supports this: the Santa Barbara Channel archaeologist Chester King (1990:65–66) identified pre–Mission period abandonment of numerous coastal Chumash villages and the consolidation of large villages into smaller "interethnic" towns. Similarly, the University of Oregon archaeologist Jon Erlandson and colleagues (Erlandson, Rick et al. 2001) have conducted decades of research on the Northern Channel Islands and argued for large-scale population declines and settlement aggregations at about the same time. These population reductions would have relaxed human predation pressure on abalone even before the start of historical otter hunting. The question remains, however, whether such declines resulted in increased abalone abundance or a coincident increase in otter populations, since the Chumash and other Native Americans also hunted sea otters prehistorically. Regardless, these Protohistoric population declines were probably only temporary, and Native American populations seem to have rebounded in the intervening years between Spanish exploration and permanent settlement.

In AD 1769, the Spanish began an aggressive colonization campaign of Alta California, instituting a pastoral economy that accelerated the breakdown of traditional Chumash society and another wave of Historic period disease epidemics. The demographic collapse of the Chumash resulted in tremendous fragmentation and reorganization of native lifeways, including an abandonment of traditional fishing, hunting, and gathering. Within 20 years, tens of thousands of Chumash and other California Indians died or were forced into slavery-like conditions (Dartt-Newton and Erlandson 2006). Between 1786 and 1803, 85 percent of the documented Chumash baptisms, conversions to Catholicism, and relocations to the missions took place. By about AD 1822, the last of the Island Chumash were removed to Spanish towns and missions on the mainland, and with their two most effective predators removed from the Northern Channel Islands almost simultaneously, abalone populations exploded in the decades that followed.

By the time the Gold Rush hit and immigrant Chinese fortune seekers arrived in San Francisco, abalone stocks throughout southern California had been largely freed from predation pressure for nearly three decades and reached unprecedented sizes and numbers. Abundant black abalone stocks were the most obvious and easiest to exploit. Unlike other abalone species in California, black abalone prefer warmer intertidal waters, and their sizes and numbers were clearly visible along rocky intertidal shorelines. Harvesting was as easy as prying the shellfish from its hold during low tide. At the same time as Chinese fisherman began to exploit other coastal resources (e.g., finfish, shellfish, seaweed), they began the first commercial harvest for black abalone.

Commercial Abalone Fishing in California

The first Chinese abalone fishing camps sprang to life in the San Francisco, Monterey, San Diego, and Baja California areas in the mid-1850s. We know few details of the history of this fishery, especially during the early years of Chinese dominance. With the state wrapped up in the frenzy of the Gold Rush, little was recorded about the specific activities of these Chinese abalone fishermen. Most of what we do know comes from scattered newspaper accounts, dock reports, annual catch records, immigration and tax files, and other historical documents.

These accounts allow us to tease out some details on the history and intensity of the fishery and the harvest techniques. The documents suggest that Chinese abalone fishermen utilized traditional skills and equipment, collected mostly black abalones along rocky intertidal shores, and processed and dried the meat for other immigrants or for overseas markets, primarily in China and Japan. An early account of this commerce was reported by the *Los Angeles Star*: "The extent and importance of our coast fisheries are not, we think, generally known or appreciated. A very large business is being carried on, in this department, along the coast and the islands in the channel, from Santa Barbara down.... We were not agreeably surprised to find the number of men and vessels engaged in our fisheries as well as the capital expended in fitting out the same.... [Chinese fishermen] are not limited to any particular kind of fishing, taking...a large quantity of [abalones]" (March 2, 1861, p. 3).

Black abalone became the focus of the Chinese abalone harvest due mainly to the ease of collection. Unlike subtidal species such as reds, pinks, and whites, blacks could be harvested along rocky shorelines at low tide with little equipment, such as the hardhat diving technology required for deep-water species. Removal of the abalone took only a basic prying device, which was slipped under the shell and used to force the abalone's foot off the rock. Prehistoric Native Americans often used a whalebone or hardwood pry bar about three to four feet long with a beveled end (Hudson and Blackburn 1982:253–255). The Chinese used long poles with a metal chisel-like implement at the end or a three-foot metal pry bar (an "ab iron"). An early *New York Times* article describes these tools and the collection of black abalone: "Every morning the hunters started out, some wading along the rocks at low tide, armed with a long pole, on the end of which was an implement like a chisel, intended for prying off the shells from the rocks" (May 16, 1897, p. 13).

After a sizable collection was accumulated, the meat was locally processed. A number of newspaper accounts illustrate Chinese processing procedures, most describing the boiling, drying, and packaging of abalone meat for shipment. This intensive process helped prevent spoilage and substantially reduced the weight and bulk. The abalone meat was removed from its shell, boiled, and laid out to dry on wooden racks (Berryman 1995: 123). When dry, the meat was packed into wooden crates or gunnysacks and shipped to the nearest port. C.L. Edholm (1913:383–384) described the process in an *Overland Monthly* article:

> There has been considerable activity among the Oriental fishermen along the California coast, who secure many tons [of abalone] for market by a long and complicated process. They are removed from the shells, salted for several days, thoroughly cooked in boiling water and dried in the sun. After they have been well dried they are again cooked, smoked for twenty-four hours, given a third boiling and once more set out on trays to dry, this time for a period of six weeks. They are then given a final bath to remove any dirt that may have accumulated, and are ready to ship to the Orient or to retail in the queer little stores of the Chinese quarters of our own cities. They can be seen there, exhibited in little glass jars, brown and uncanny looking articles, which are apparently as tough as sole leather, but they are very highly prized as a toothsome morsel.

Processing techniques likely varied based on the habits of the fishermen, their gathering locations (mainland versus islands), and any constraints of time or equipment. Another account, from the *San Diego Union*, for example, recorded slightly different processing procedures:

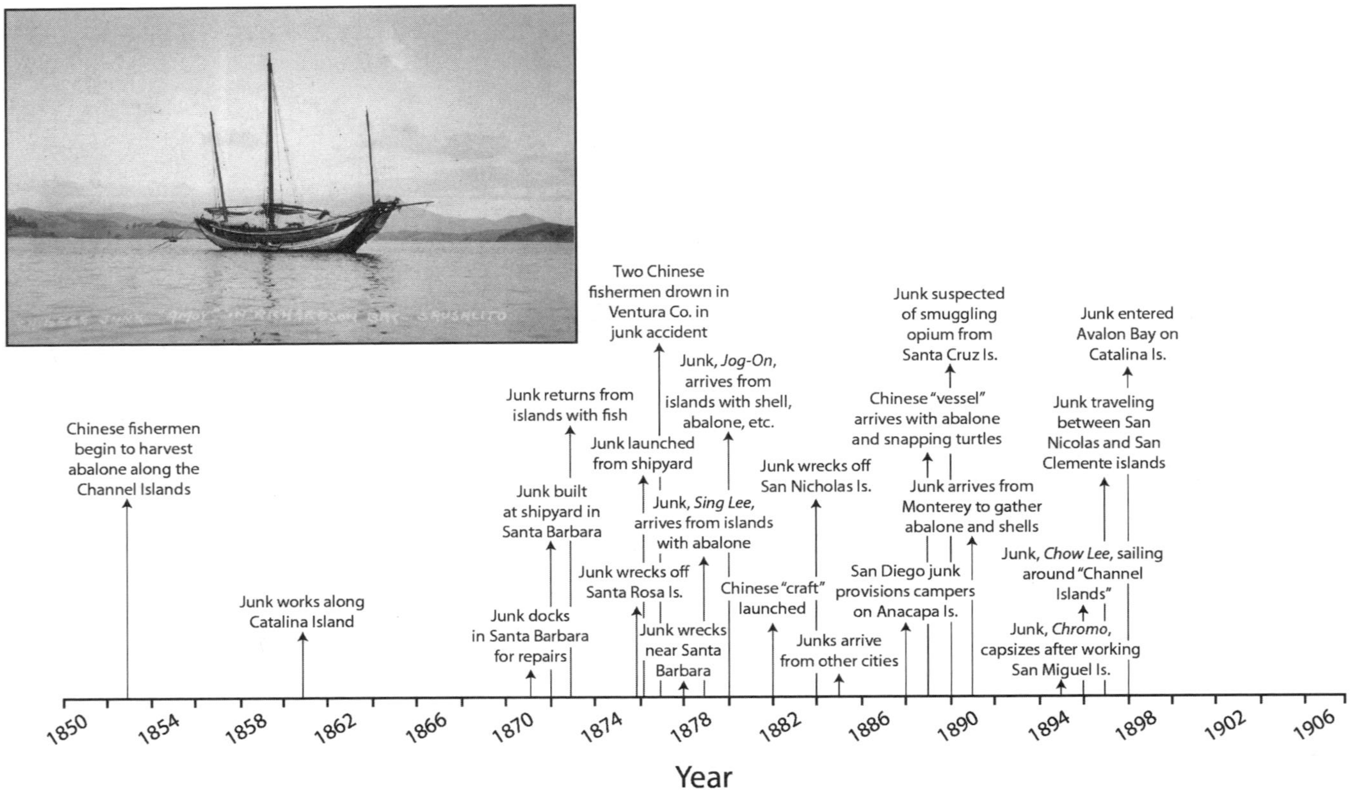

FIGURE 5.3. Timeline of Chinese junks mentioned in various historical newspaper accounts from Santa Barbara. *Inset:* Historical photo of the Chinese junk *Amoy* in Richardson Bay in Sausalito, California. (Open access image via the Bancroft Library, University of California, Berkeley, accession number 121)

> The abalone is found attached to rocks; and in extreme low tide in greater quantities than at any other time. The fish, covered by dense shell, adheres to the rock as tightly as if glued to it, and is cut loose with a sharp instrument carried by the Chinamen. After filling a large bag with meat, which is removed from the shell, it is carried to the place chosen to dry it...it receives a good pounding. After this beating, the meat is thrown into a large kettle and boiled for a short time, then it is spread out to dry in the sun. After a thorough drying, it is nicely packed in strong sacks and shipped to San Diego, to be reshipped from here to Chinese merchants in San Francisco. The meat commands in that city from five to six cents per pound, and is used exclusively by the Chinese. A considerable quantity is shipped to China, where it is regarded as a great luxury, being only used by the better class of people in the country. (October 5, 1871, p. 3)

At a minimum, Chinese abalone fishermen required hearths and boiling pots, wood for drying racks and fuel, and campsites substantial enough to maintain collecting and processing operations for weeks to months at a time.

According to newspaper accounts, Chinese fishermen used traditional watercraft, junks and possibly sampans, to access prime abalone habitat, and junks also transported fishermen and product to and from the Channel Islands. It is unclear exactly when junks were first operated in southern California, but the earliest newspaper account comes from 1861, suggesting that by the late 1850s to early 1860s, Chinese junks were employed in the fishing industry. The *Santa Barbara Daily Press* reported on September 8, 1879, for example, "The *Sing Ling*, a genuine Chinese junk, arrived yesterday from Santa Barbara Islands with a cargo of abalone shells." Figure 5.3 illustrates a sample of the Chinese junks observed operating in the Santa Barbara Channel region as mentioned in various historical newspaper accounts. Often these junks were owned by wealthy Chinese merchants in port cities such as Santa Barbara or San Diego and managed as partnerships with other fishermen or leased to operators in need of a vessel (see Bentz and Schwemmer 2002:143–144).

It may have been even more common for Chinese fishermen to move to and from the Channel Islands aboard Euro-American-owned vessels.

Historical newspaper accounts suggest that Chinese fishermen frequently contracted with sloop and schooner captains between 1873 and 1911. A variety of such vessels held periodic or continuous contracts for the transportation of men, gear, and abalone catches from the Santa Barbara mainland to specific islands such as Santa Rosa or Santa Cruz.

Smaller, less seaworthy sampans are rarely, if at all, mentioned in historical documents. These vessels probably were transported to abalone gathering locations by larger schooners, sloops, and junks, and fishermen, supplies, and sampans were often dropped off to harvest a local coastline for an extended period of time until junk and other vessel operators returned to bring both the crew and cargo back to port. Newspaper accounts suggest this to be the case especially with abalone operations on the Channel Islands. A sobering account of the inherent dangers of the system was reported by Santa Barbara's *Morning Press* on March 27, 1900:

> On bleak San Nicolas, the most distant and desolate of the Santa Barbara group of islands, three Chinamen were for the past three months forced sufferers from hunger, and one is dead from the results of the prolonged fast—died from starvation, from the lack of sufficient and proper food. The two were rescued just in time to save their lives, for they were weak and failing rapidly when discovered. Two weeks' delay, at least, would have meant their death. Somewhere up or down the coast is an unknown sloop that was three months ago manned by unknown sailors, and to them is chargeable this suffering and death. While the three Chinamen were absent, hunting abalones on the opposite side of the island, these sailors, who had been catching crawfish on San Nicolas, raided the Chinamens' [*sic*] camp, stole their rice and 1500 pounds of abalone meat, and put out to sea. When the Chinamen returned they found their provisions gone, but they were helpless, having no boat large enough to brave the sea that rages between them and the mainland nearly 100 miles away.

The two remaining Chinese fishermen were ultimately rescued after their employer in Santa Barbara chartered a vessel to pick them up, unaware of their peril. The article offers intriguing insights into their foraging strategies, stating that the fishermen had "no boat large enough" to make the trip back to Santa Barbara. The fishermen, then, probably had some type of watercraft and were likely using sampans to forage the island coastline for months at a time from a centralized basecamp where supplies and cargo were stored.

Other historical newspaper accounts offer more routine and less dramatic details that support the notion that Chinese fishermen occupied the islands for extended abalone fishing expeditions, established long-term basecamps, and awaited return to the mainland aboard junks or schooners. Little more than six months after reporting the drama on San Nicolas Island, for example, the *Morning Press* (October 4, 1890) reported business as usual: "The junk *Acme* arrived yesterday from San Miguel island with a cargo of abalone shells from the Chinese fishing camps." This and other accounts suggest that junks and Euro-American vessels regularly resupplied fishermen and transported cargo. Most likely, this was done from known basecamps near safe anchorages at scheduled times.

Further evidence can be gleaned from the immigration files of nineteenth-century Chinese abalone fishermen housed at the National Archives in Perris, California. Lee Tong's case file (Case File 277, 1897), for example, describes his partnership with Ah Poy, a Chinese merchant who operated a business collecting abalone from Santa Rosa and the other Channel Islands. Tong's fishing outfit included "boats, tents, kettles, crow bars and such utensils." Another Chinese fisherman, Tai Kin (Case File 269, 1897), is described as a foreman "over catching abalone fish" who worked for a company whose fishing equipment included boats, trays, fishing tackle, and drying racks. Both of these files suggest that boats (sampans) were an integral part of island abalone fishing.

This abalone foraging pattern, in which junks largely operated as transport vehicles, was not universal and may have been more common on the Channel Islands than along the mainland. Historical accounts indicate that junks regularly left their home ports for extended periods to forage abalone up and down the Alta and Baja California coastlines. The *Los Angeles Times* reported on June 6, 1896, for example, "The Chinese junk *Chow Lee* sailed into port Wednesday, after being out seventy-nine days. She had a cargo of abalone meats, abalone shells and dried squids, all of which smelled to heaven with odors unspeakable."

For the first decade or so of the commercial abalone industry, the meat was the target, dried and shipped to domestic and foreign markets. Abalone shells were discarded in piles along the coastline; prices per shell were too low to bother shipping them to market. That is not to say they had no value at all. Historical documents describe the creating and selling of rings, necklaces, and other forms of personal ornamentation from abalone shell as early as the late 1840s. The historian Hubert Bancroft (1890: 97) wrote, "The manufacture of jewelry was fostered toward the close of the forties, by miners who desired specimens polished, set, or made into chains and rings. Abalone shell and quartz soon became a specialty sought by all visitors." The shell market in these early years was probably small and localized. Bancroft (1890) never reported the prices per shell or who was involved in the industry.

Beginning in the mid-1860s, however, commercial interest in lacquerware, shell ornamentation and inlay, and lime for construction cement skyrocketed, and demand for abalone shell grew exponentially (Bentz 1996:16). Chinese abalone fisherman began shipping both the dried meat and shells to market, and returned to old abalone processing camps to retrieve discarded shells. Shells were shipped out of San Francisco to the East Coast, Europe, and China and manufactured into buttons, jewelry, polished mantel ornaments, and other curios. An early description of the polishing process was recorded by George Brown Goode (1887:596): "Great interest is taken by the Chinese in abalone fishing. The shells of the young of the species *Haliotis splendens* are treated with a solution of hydrochloric acid. The shells of the adult individuals of the same species and those of another species are ground down on stones by hand. Steam grinding, from the rapidity of motion, wears holes through the shells, unless the operator is extremely careful. The shells, when ground down, are varnished."

Market prices for abalone shells fluctuated, at times making them more valuable than abalone meat. As early as 1870, for example, high prices were paid for the shells of various abalone species. In a California Fish and Game Commission report, Charles Lincoln Edwards (1913:8–9) recorded the prices of abalone shell in Avalon on Catalina Island: "The meat sold for five cents a pound, the green shells brought eighty dollars a ton. At the present time, the green shells are sold at one hundred and twenty-five to one hundred and eight dollars a ton, the black, at eight to one hundred dollars a ton, and the red at forty to seventy-five dollars a ton. The black shells, with especially good pearly centers, bring from three hundred to five hundred dollars a ton. Owing to the increasing scarcity of good green shells, there is a growing tendency to use the centers of the red shells for jewelry." By 1888, abalone meat fetched 5 cents per pound and shells 12 cents per pound (Bentz 1996:16). The shell market was so volatile, however, that shortly after a period of relatively high market prices, a U.S. Commission of Fish and Fisheries Report recounted, "The unremunerative prices now obtained for the shells deter white men from engaging in the fishery for them" (Collins 1888:45).

Historical records can take us only so far in understanding the history of Chinese abalone fishing and the birth of commercial abalone exploitation in California. These documents lack details about the nature and size of Chinese camps and what life was like for the fishermen. Very few firsthand accounts are available, and most descriptions of the industry are related from afar by secondhand sources and often colored by the racist attitudes toward Asian immigrants that were so prevalent at the time. Fortunately, we can turn to the archaeological or material culture record to glean additional insights and critically evaluate the written sources.

The Material Remains of Historical Abalone Fishing in California

The major obstacle to better understanding the history of Chinese abalone fishing in southern California is a taphonomic one (see Chapter 3). The archaeological signatures of most temporary abalone fishing camps on the mainland have been lost to a variety of cultural and natural transformation processes, the two major categories of taphonomic forces. These transformation processes include all the conditions and events that affect archaeological materials from the time of deposition to the time of recovery. Natural (erosion, earthquakes, animal burrowing, decay, etc.) and cultural (the result of past or present human behavior) transformation processes have been particularly destructive on these site types and have destroyed much of the mainland record. The reasons for this are complex, but generally, historical abalone camps are found embedded in unstable dune sands and Holocene soils, along windy, exposed, and heavily eroded shorelines. Due to natural processes of undercutting, shoreline retreat, and wave action, these sites have rapidly destabilized along the mainland.

Perhaps the most destructive force, however, has been the exploding human population along the California coast. Much of the construction and development that has fundamentally altered the California coastline to accommodate the influx of people began before the implementation of state and federal laws that now protect archaeological resources in North America. Chances are, most nineteenth-century Chinese abalone processing sites along the mainland were destroyed to make way for houses, roads, and strip malls. Fortunately, a variety of state and federal laws now protect these and other archaeological sites, and cultural resource managers are tasked with ensuring their preservation. Additionally, nineteenth-century abalone fishermen themselves affected the archaeological preservation of the early abalone fishery. After the market for abalone shells developed, fishermen shipped the most obvious indication of an abalone fishing camp—abalone shells—to mainland markets. Fishermen also mined abalone shells from pre–shell market sites, dramatically reducing their archaeological visibility.

Archaeological excavations at more permanent mainland Chinese mining camps, stores, and Chinatowns have helped reveal fascinating details about the lives and activities of nineteenth-century Chinese immigrants throughout the American West (e.g., Barlow and Richardson 1979; Baxter and Allen 2002; Costello et al. 2008; Greenwood 1996; Lister and Lister 1989; Voss 2005; Wegars 1993). Excavations at these types of sites often produce a diverse array of artifacts, structures, and living surfaces. The archaeologist Roberta Greenwood (1996), for example, excavated a portion of downtown Los Angeles's old Chinatown, occupied between the 1880s and 1933. Greenwood (1996) and colleagues unearthed historical trash pits, hearths, structural foundations, living surfaces, privies, and other archaeological features. These excavations produced hundreds of decorated and utilitarian ceramics, eating utensils, ornaments, beads, toys, gambling pieces, coins, medicines, personal items, and much more.

Excavations also produced evidence for black abalone consumption in mainland Chinese households. Greenwood (1996:128) reports the presence of small amounts of black abalone shell at the Los Angeles Chinatown, as does a 1901 food study on the diet of Chinese households in San Francisco (see Gust 1993:192; Jaffa 1901). Subsistence studies from archaeological deposits probably underestimate the importance of black abalone to nineteenth-century Chinese diets, however, as most of the abalone consumed likely entered households as dried abalone meat. Its consumption probably left a minimal distinctive archaeological signature. In addition to the ravages of mainland Chinese fishing sites by taphonomic processes, nineteenth-century Chinese abalone fishing and consumption remains understudied.

Fortunately, a substantial record of the early abalone fishery still exists, preserved along California's offshore islands. Largely protected from historical development by Channel Islands National Park, the U.S. military, the Nature Conservancy, and the Catalina Island Conservancy, the Channel Islands contain a fascinating archaeological record of historical abalone fishing that can supplement catch records and occasional newspaper accounts. Even the sites still preserved on the offshore islands, however, are probably just a small fraction of a once robust and extensive archaeological record. Many of the sites that have been identified are located along windswept shorelines embedded in unstable dune sands, where they have been rapidly eroding for some one hundred fifty years. In some island locations, historical abalone processing sites have been identified in the midst of sea mammal haulouts and rookeries (see Chapter 6), where their archaeological remains have been regularly crushed and reshuffled.

When preserved, Chinese abalone camps are highly distinctive and easily differentiated from Native American shell middens because of the almost exclusive presence of black abalone shells and the relatively large size of the shells (see Braje and Rick 2015). It is important to keep in mind as well that these island sites probably do not represent the entire history of commercial abalone exploitation. Most were likely created during the earliest years of the fishery up until the mid-1860s, before the shells were shipped to mainland markets. After this point, fishermen exported the most abundant and robust indicator of their activities, leaving very little archaeological signature. Even abalone piles representing early commercial fishing are probably only the basal remnants of once massive shell piles that were later mined to feed the ornamental and inlay markets of later decades. It is possible, however, that some historical abalone shell piles and processing camps represent intervals when the market for shells crashed, making their shipment to the mainland uneconomical.

At most sites, piles of black abalone shell are the only archaeological signatures. Finding any historical artifact is a rare occurrence, and temporally diagnostic artifacts are exceptionally uncommon. Some sites, however, have produced Chinese pottery sherds, cartridge casings, metal and glass fragments, and other historical debris. An even smaller fraction of sites contain hearth features, presumably where abalone meat was

FIGURE 5.4. Examples of Chinese brown-glazed stoneware vessels occasionally identified at nineteenth-century Chinese abalone fishing sites on the Northern Channel Islands: (*a*) liquor bottle, (*b*) barrel jar, (*c*) spouted jar, (*d*) straight-sided jars, (*e*) globular jar, (*f*) wide-mouth jars. (Reprinted by permission from J. Costello)

boiled before shipment to the mainland. In some cases, sites can be confidently identified as "Chinese" historical abalone processing camps with material culture because of the presence of distinctive "hairpin-shaped" hearths (commonly found at known nineteenth-century Chinese sites on the mainland; see Chapter 6), brown-glazed stoneware or other Chinese ceramics, opium paraphernalia, or other temporally diagnostic material culture.

The most distinctive and ubiquitous indicator of Chinese occupation on the Channel Islands is the presence of Asian ceramics (Figure 5.4), produced in China for domestic or overseas markets (Williams 2011:365). These ceramic vessels were brought by Chinese fishermen out to the islands and contained not only their traditional foods but also a cultural and symbolic connection to their homeland. Brown-glazed stoneware vessels are the most common type found in nineteenth-century historical archaeological sites and served utilitarian purposes as containers for food and drink:

> Small jars with a spout on one side were used to hold food items such as soy sauce, black vinegar, peanut oil, and liquor. Small

> bulbous bottles with a flaring rim held both distilled and fermented alcoholic beverages. Wide-mouth jars came in various sizes and were used for preserved foodstuffs such as tofu, beans, and bean paste, pickled turnips, cabbage, and shrimp paste. Large, globular, and small-mouthed jars were usually used for liquids, hard liquor, soy, and oils and were commonly reused for water storage or sometimes as pickling crocks. The smaller sized jars with straight sides were for household use and contained herbs, preserved tofu, and other cooking substances. They could also be used to steam food. Barrel jars are very large and straight-sided. They were used for shipping and storage of large quantities of foodstuffs, such as rice, grains, soybeans, sheet sugar, and sticky rice powder. These large jars were also used to ship the bones of the dead back to China. (Yang and Hellman 1998, as cited in Costello et al. 1999:252)

These vessels came in a variety of shapes and sizes and usually were glazed with some shade of light to dark brown. Occasionally, however, stonewares were produced with a variety of other colored glazes, such as green shades on hexagonal "ginger" jars or "liquor bottle" shapes with black to blue-black hues (Muller 1987). Archaeologists have commonly classified these vessels, based on their perceived function, as soy sauce jars, large shipping jars, globular jars, or ginger jars, to name a few (see Greenwood 1996).

Less common than utilitarian storage and transportation stoneware vessels are Asian tablewares, employed as serving vessels, dishes, and cups (Figure 5.5). Produced in China, these light-colored porcelain vessels are commonly identified by their decorative motifs as bamboo, four flowers, double happiness, or celadon (see Greenwood 1996; Muller 1987). Bamboo vessels are rice bowls that tend to be thick-walled, gray, and of coarse texture, with hand-painted cobalt blue "plant forms, circles, and representations interpreted as dragonflies" under the glaze (Greenwood 1996:54). Four-flowers decorations are commonly found on a variety of table vessels, from serving bowls to rice bowls to condiment dishes and plates (Greenwood 1993:60–70). Double-happiness ware types are found as a variety of tablewares and were among the earliest and cheapest types of Chinese porcelains. These vessels are twice decorated with the Chinese character *hsi* (happiness). Celadon wares, most commonly found as small teacups,

FIGURE 5.5. Examples of Chinese bamboo (*back row*) and celadon (*front right*) bowls, and a round ginger jar (*front left*). (Reprinted by permission from J. Costello)

have a thick, wintergreen-color glaze, and (often) blue character marks on their bases. Double-happiness bowls are found only at sites that predate 1870, while bamboo and celadon are the two most commonly recovered types from California Chinatown sites dating from the later nineteenth to early twentieth centuries (Greenwood 1996:70).

Even without Asian ceramics or other artifacts, the characterization of historical abalone processing camps as "Chinese" is probably reasonable, although it is possible that members of other ethnic groups contributed to the formation of some sites. Perhaps the strongest support comes from the fact that nearly every historical abalone camp identified on the Channel Islands contains one or more piles of abalone, almost exclusively composed of black abalone shell. As we see in Chapter 7, later groups of abalone fishermen, including Japanese divers and Euro-Americans, who entered the industry at the turn of the twentieth century, were collecting not only black abalone but also red, green, pink, and white abalone, whereas Chinese fishermen exclusively focused their harvest on intertidal blacks.

Of the eight offshore Channel Islands, seven have been archaeologically surveyed for historical abalone processing camps. Although newspaper accounts describe Chinese abalone fishing on all of the islands, Catalina Island has never been systematically surveyed for these site types, and despite extensive survey, no historical abalone camps have been identified on Santa Barbara and Anacapa Islands. Even without archaeological

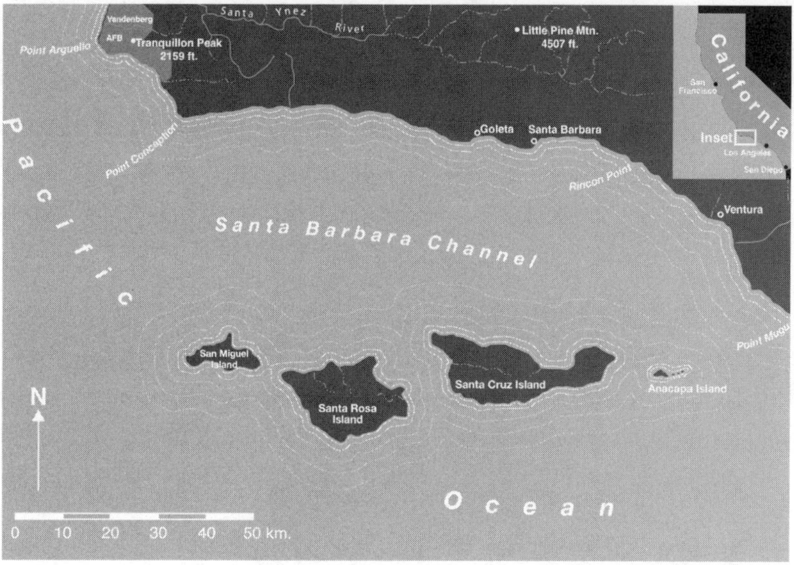

FIGURE 5.6. Map of the southern California Bight and the Northern Channel Islands.

evidence, however, it is safe to assume that these islands were fished by Chinese abalone hunters and that natural erosion and later human activities (ranching, navy activity, etc.) destroyed the material remains. For example, in a late-nineteenth-century article titled "The Santa Barbara Islands," Homer P. Earle (1896:227) wrote, "Anacapa is uninhabited, unless one counts campers and the Chinamen who spend part of the year in a tent amid piles of iridescent abalone shells and square yards of the evicted tenants, drying in the sun, to be shipped to Chinatown." The remaining Southern (San Clemente and San Nicolas) and Northern (San Miguel, Santa Rosa, and Santa Cruz) Channel Islands all contain an extensive record of commercial abalone fishing (Braje, Erlandson, and Rick 2007). These data offer a window into the lives and activities of Chinese abalone fishermen in southern California.

Since my fieldwork and archival research has centered on the Northern Channel Islands, I focus on the material record of historical abalone camps on these islands (Figure 5.6). It is important to note, however, that San Clemente and San Nicolas Islands, in the Southern Channel Islands, contain a fascinating record of historical abalone fishing, and considerable archaeological work has carefully documented this record. On San Clemente Island, archaeological surveys beginning in the 1950s documented 43 such sites, hundreds of historical artifacts, and numerous

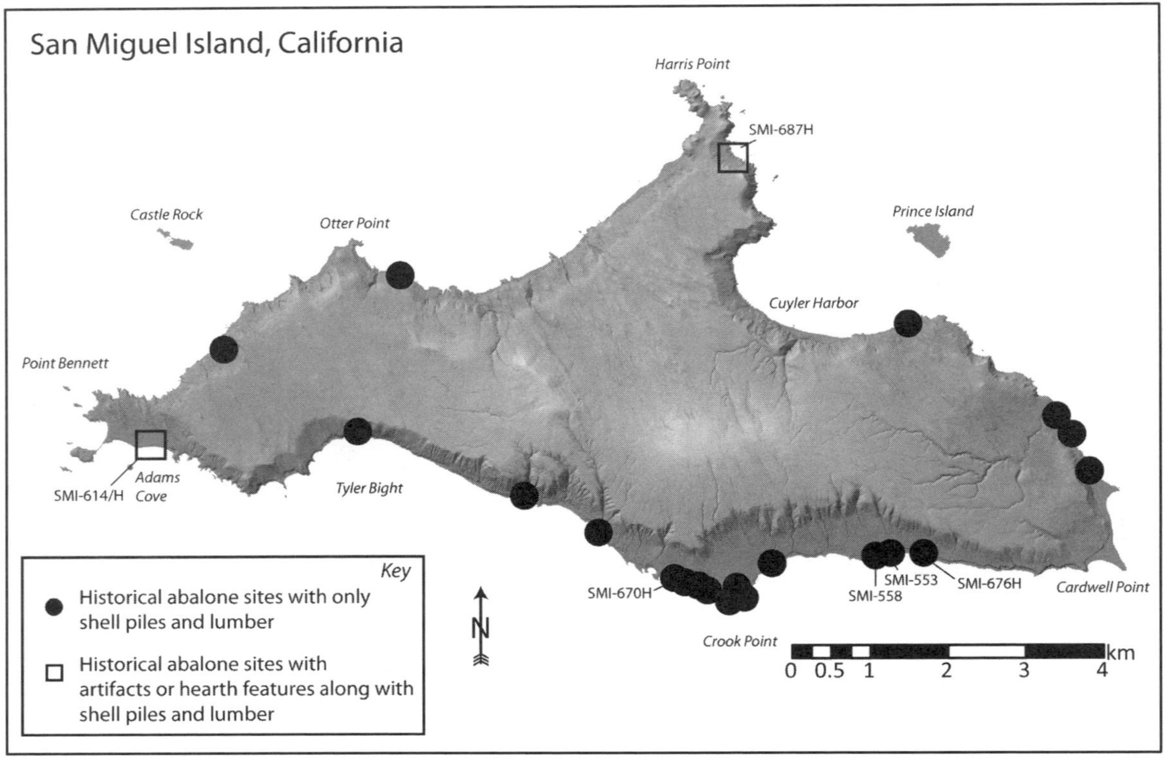

FIGURE 5.7. Map of the 22 historical abalone fishing sites identified on San Miguel Island. One of the sites, CA-SMI-614/H, contains diverse and diagnostic historical artifacts and features. *Note:* Site locations are approximate due to National Park Service confidentiality concerns.

hearth features, shelters, and windbreaks (see Axford 1984, 1987; Berryman 1995; Braje, Erlandson, and Rick 2007; McKusick and Warren 1959). On smaller San Nicolas Island, 26 historical abalone fishing camps have been identified; most consist only of black abalone shell piles, but a few produced temporally distinctive Chinese ceramics and hearths along with other nondiagnostic historical artifacts such as glass bottle fragments, amorphous metal fragments, and lumber (Berryman 1995; Krautkramer 2002; Reinman and Lauter 1984; Schwartz and Rossbach 1993). A review of the available archaeological data across the Northern Channel Islands suggests slight differences in harvesting techniques and activities but a fishing industry that was generally similar across the channel.

San Miguel Island

In 1982, the University of California–Santa Barbara archaeologist Michael Glassow recorded the first historical abalone fishing sites on the Northern Channel Islands. With research agendas focused on ancient (Native American) sites, earlier researchers tended to bypass historical cultural resources in favor of identifying, dating, and sampling prehistoric sites created by the Island Chumash and their ancestors. While surveying two quarter-mile-wide transects running north-south across San Miguel Island, Glassow (1982) identified 22 previously unrecorded archaeological sites. Two of these sites, CA-SMI-553 and -558, located at the sea cliff exposure on the south coast, were described as lenses of mostly whole black abalone shells. Based on the site stratigraphy and their similarity to historical shell middens identified on San Clemente and San Nicolas Islands, Glassow (1982) suggested that they were historical abalone collection and processing locations.

More than two decades later, in the summer of 2003, I organized a team of archaeologists to conduct the first systematic archaeological survey on the Northern Channel Islands with the specific purpose of identifying historical abalone fishing locations (Braje and Erlandson 2006). This shoreline survey of San Miguel Island focused on sea cliff exposures and the supratidal region and documented 20 unrecorded historical abalone sites, bringing the total number of these site types to 22 (Figure 5.7) (Braje 2010:99–100). Additional sites may exist on offshore islands or rocks (i.e., Prince Island and Castle Rock), but they have never been surveyed and are not mentioned in historical newspaper accounts or other documentation.

FIGURE 5.8. Typical historical abalone fishing sites identified on San Miguel Island and the other Northern Channel Islands. Most of these sites contain only black abalone shell clusters embedded in historical dune sands along sea cliff exposures near rocky intertidal headlands. (*A*) A heavily eroded and relatively large cluster of black abalone shells on Santa Rosa Island at CA-SRI-798H. (*B*) A large and heavily eroded cluster of black abalone shells eroding from the high supratidal zone at CA-SMI-667H. (*C*) One of several loci of black abalone shell piles at CA-SMI-663H; a navy range pole can be seen in the upper left corner. (*D*) The eroding sea cliff exposure at CA-SMI-558 on San Miguel Island, one of the first two historical abalone fishing sites ever recorded on the Northern Channel Islands. (Composite and images by author)

In addition, very small piles of large black abalone shells have been identified on the southwest end of the island near the coastline, but they have not been recorded as unique sites and may be best understood as part of the large historical abalone basecamp in the area (CA-SMI-614/H; see Chapter 6).

Despite extensive sea cliff exposures and careful surface reconnaissance, few historical artifacts or features were found at San Miguel abalone camps, and most sites contained only piles of large, mostly whole black abalone shell (Figure 5.8). Scattered pieces of lumber, possible remnants of drying racks, were found at three sites, and two sites contained features other than shell middens. One site, CA-SMI-670H, located just west of Crook Point, included two large, beach-rolled stone slabs near a pile of discarded abalone shells. The function of these rocks is uncertain, but there was no evidence of burning, so they were probably not related to hearth features and may have served as sitting stones or drying slabs. At the lone historical abalone fishing camp on San Miguel's north-central coast, CA-SMI-687H, near Harris Point, a diagnostic Chinese hairpin-shaped hearth was identified at the sea cliff exposure in association with a small black abalone shell pile. Although no historical artifacts have been identified, the feature is constructed from angular bedrock cobbles, and oxidation (reddening) suggests that it served as a hearth, once used by Chinese fishermen to boil abalone meat. This site and the hearth have yet to be excavated, so it may be that historical artifacts are buried below the surface.

The only other historical abalone fishing camp to produce artifacts or features on San Miguel is CA-SMI-614/H, located on the far west end of the island at Point Bennett. This site, first identified by the National Marine Fisheries Service biologist Robert DeLong, has produced the greatest density and diversity of Chinese (and later) historical abalone fishing artifacts and features ever recorded on the Channel Islands. Even though the site has been heavily disturbed by sea mammal activity and coastal erosion, it has produced a truly astonishing set of historical artifacts and features, including Asian ceramics, abalone buttons, ivory rings, gun cartridge casings, glass and metal fragments, and rock hearths, to name a few. No single site on the Channel Islands has done more to help us better understand the activities and lives of Chinese abalone fishermen in southern California. In Chapter 6, I explore the Point Bennett historical site (CA-SMI-614/H) in greater detail.

In addition to our survey, we conducted the first subsurface excavations of historical abalone fishing camps on the Northern Channel Islands, carrying out small-scale archaeological testing of black abalone shell piles at CA-SMI-558 and -676. From eroding exposures at the sea cliff, a 100-liter bulk sample and a 2.0-×-1.0-m excavation unit was excavated at CA-SMI-558 along with a 45-liter bulk sample and a 0.5-×-0.5-m unit at CA-SMI-676H (for details, see Braje, Erlandson, and Rick 2007). These excavations were designed to sample the site constituents, provide a measure of the density and average sizes of the black abalone shell (see Chapter 7), and identify diagnostic subsurface artifacts and features. Unfortunately, no such diagnostic artifacts or features were identified at either site. At both CA-SMI-558 and -676H, black abalone dominated the deposits, comprising over 99 percent of the recovered constituents by weight. Besides black abalone shell and tiny amounts of other shellfish and fish bones, the only materials identified at either locality were a cluster of six rocks embedded in dune sand overlying a pile of black abalone shells at CA-SMI-558. While these rocks were clearly transported by humans, we can only speculate as to their purpose, but likely they were employed as part of abalone drying racks.

Santa Rosa Island

Don Morris, former Channel Islands National Park archaeologist, documented the first historical abalone camps on Santa Rosa Island. As part of his systematic shoreline survey for archaeological resources in the early 1990s, Morris (1994) and his team identified 12 of these site types, nearly all clustered along the southern and western shores of the island. Most of the sites consisted of concentrated piles of large black abalone shells embedded in dune soils near abundant rocky shorelines. Since Morris and his crew were focused on sites that could be identified as Chinese, many of the sites they recorded contained historical artifacts or features in addition to black abalone piles and lumber planks, although four were distinguished based solely on these features (CA-SRI-31/H, -109, -359/H, and -472/H). Eight sites produced not only black abalone shell piles but also hearth features and/or heavily eroded artifactual remains, including fragments of stoneware and porcelain pottery, opium paraphernalia, cartridge casings, amorphous metal, and bottle glass (Bentz 1996:37; Don Morris, personal communication 2005).

CA-SRI-559, located immediately to the west of Carrington Point on the north-central portion of the island, contained a black abalone pile along with several pieces of wood planking (possible drying ranks) and heavily eroded iron rods, one anchored in the sandstone substrate. On the far southwestern coast, near South Point, Morris (1994) identified a black abalone shell pile and two rock features (CA-SRI-353). One was a possible hearth feature, recorded as an open circular structure 4 m in diameter and constructed of sandstone boulders. The second consisted of a 2-m-long sandstone wall, with several wooden stakes driven into the structure, perhaps the remnants of a drying rack base. Morris identified

Chinese brown stoneware ceramic sherds at CA-SRI-40, just east of Brockway Point, and CA-SRI-95, in the Cluster Point area, but both sites have since eroded into the ocean and nothing remains to further document Chinese activities in these areas.

Morris and his team also identified a large Chinese site, CA-SRI-91-75 (a locality that remains officially unrecorded, although the site is mapped with a temporary number), on the northwest coast of the island at the western tip of a large rocky point. The site has produced five large black abalone piles, a badly eroding hairpin-shaped hearth with abundant charcoal and amorphous metal fragments scattered near its mouth and interior, approximately 100 scattered lumber pieces sharpened into points at one end, and nine brown stoneware jar fragments, six bamboo bowl fragments, and five celadon bowl fragments. Blowing sands and considerable erosion have heavily affected the preservation of the site, and continued monitoring and archaeological investigations will be necessary to document the size, extent, and significance of this locality. Initial mapping and surface collections suggest that this location was a large basecamp where black abalone were collected, boiled, and dried.

The largest and most diverse site identified by Morris, CA-SRI-471, is located on the south-central coast near South Point at Johnsons Lee. This complex and heavily eroded site can be definitely connected to the Chinese abalone fishing industry and has produced one of the two largest and most diverse densities of Chinese artifacts ever found on the Channel Islands. I discuss recent archaeological investigations and important insights about the Chinese abalone fishing period that can be gleaned from this locality in the next chapter.

To contextualize the remaining two Chinese sites containing diverse artifactual remains identified by Morris (1994), we must begin in Greenock, Scotland, in 1883, when a four-masted, iron barque British clipper ship, the *Goldenhorn*, was built by Russell and Co. for J.R. de Wolf and Son (Lubbock 1927:244). At the time, the *Goldenhorn* was a hallmark of technological innovation in working cargo ships. Though it was not the fastest vessel, the *Goldenhorn* offered an ideal compromise among speed, cargo capacity, and ease of operation, and for nearly a decade, it successfully hauled freight around the world from distant ports in Southeast Asia, Europe, and the New World.

FIGURE 5.9. CA-SRI-241/H, a Chinese archaeological site associated with the 1892 *Goldenhorn* wreck: (*A*) large black abalone shell pile near the sea cliff exposure; (*B*) historical lumber likely gathered from the wreckage of the *Goldenhorn*; (*C*) large coal pile eroding from the sea cliff exposure. (Composite and images by author, June 2012)

In its final voyage, the *Goldenhorn* carried a crew of 28 men and 2,808 tons of coal from New South Wales, Australia, to San Pedro, California, for use by the Southern Pacific Railroad (Wheeler and Kallman 1984:37–38). On the evening of September 12, 1892, caught in heavy fog and shifting currents, the *Goldenhorn* wrecked on offshore rocks along the southwestern coast of Santa Rosa Island. The crew quickly boarded two lifeboats and rowed for 24 hours, across the channel, to Santa Barbara, where locals kindly supplied food, clothing, and shelter (Wheeler and Kallman 1984:37). The only loss of life was the ship's cat.

Today the remains of the *Goldenhorn* can be found scattered in relatively shallow waters off the southwest coast of Santa Rosa Island, recorded and protected as a submerged historical archaeological site (CA-SRI-368). As part of Morris's (1994) coastline survey, his team identified two historical abalone fishing campsites along the adjacent shoreline from the *Goldenhorn* wreckage, CA-SRI-238/H and -241/H (Figure 5.9). CA-SRI-238/H contains the remnants of a wooden structure (possibly a shelter), a black abalone pile, and a few coal pieces tucked away in the bottom of a small gully. This ephemeral structure may be the remnants of a Chinese camp built from wooden shipwreck pieces, possibly hatch covers, salvaged from the *Goldenhorn*. Chinese fishermen may have established a short-term camp at this locality, out of the wind and elements, as they harvested black abalone from the abundant rocky intertidal zone along this stretch of coastline. CA-SRI-241/H was likely the center of abalone processing by Chinese fishermen, since it is located along the sea cliff exposure and contains a large black abalone pile and an associated pile of coal. The coal would have provided excellent fuel for boiling abalone at CA-SRI-241/H or feeding a domestic hearth at CA-SRI-238/H. Unfortunately, no diagnostic artifacts or hearth features have been identified, although this evidence may have been covered by blowing dune sands or destroyed by coastal erosion. A local newspaper account places Chinese abalone fishermen somewhere on the island five weeks after the wreck: "The schooner Santa Rosa left for Santa Rosa Island, with a crew of Chinamen who have been engaged in securing a cargo of abalone shells" (*Santa Barbara Morning Press*, October 19, 1892), lending support to such an interpretation.

Beginning in 2012, I directed a team of archaeologists from San Diego State University in a coastline survey of Santa Rosa Island to locate additional historical abalone fishing sites and to revisit those identified by Morris (1994). One of the primary goals was to identify historical abalone fishing camps left unrecorded due to the lack of historical artifacts or features other than black abalone shell piles (Braje and Bentz 2015). Although small stretches (e.g., Ford Point to East Point) were not systematically surveyed for historical sites due to the rugged and dangerous nature of the coastline, and some historical abalone fishing sites may have been destroyed by coastal erosion or other taphonomic processes, our survey achieved a nearly comprehensive coverage of the island coastline. All told, we found 14 new historical abalone processing sites (bringing the island total to 26 historical Chinese abalone fishing sites), ranging from small, heavily eroded clusters of large, mostly whole black abalone shell to more substantial basecamps with hearth features and historical artifacts (Figure 5.10). The vast majority of sites (83 percent) were found on the northwest and southwest coastline between Brockway Point to Sandy Point and Sandy Point to Johnsons Lee.

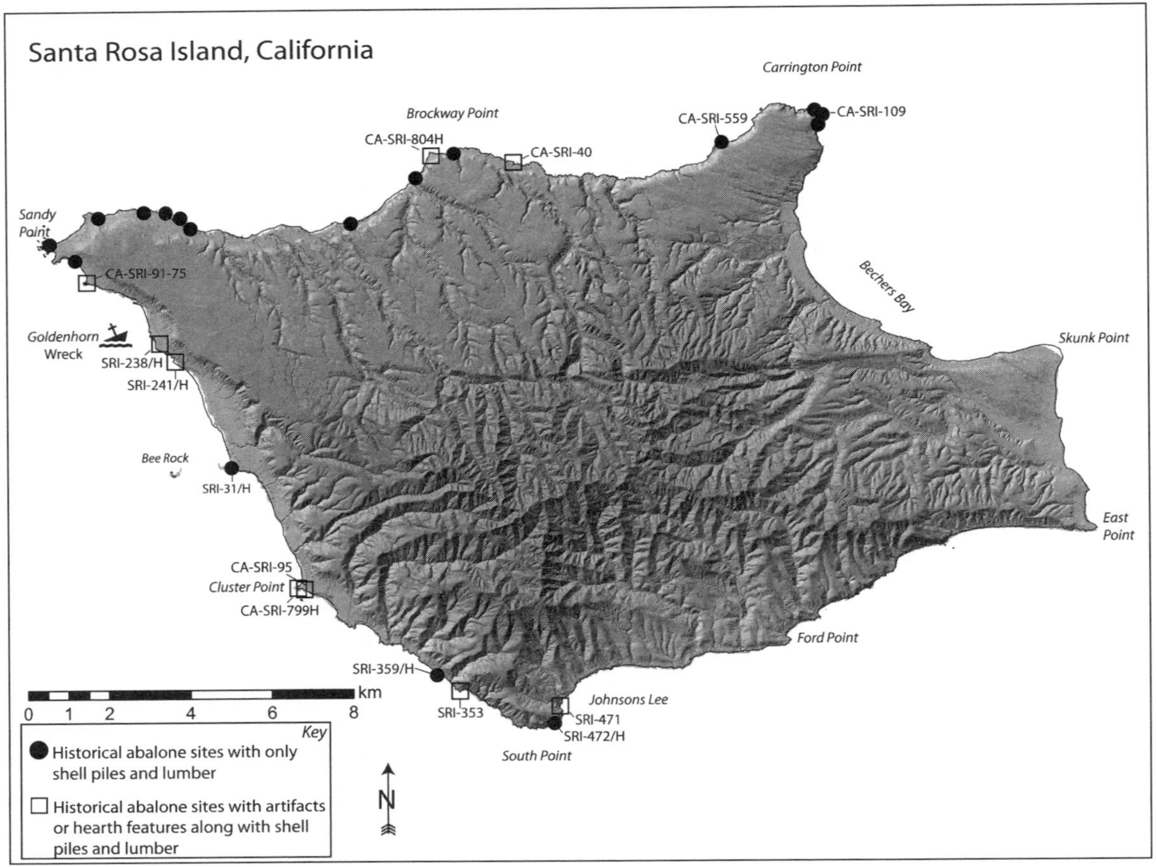

FIGURE 5.10. Map of the 26 historical abalone fishing sites identified on Santa Rosa Island. Six of the sites contain diverse or distinctive historical features and/or artifacts. *Note:* Site locations are approximate due to National Park Service confidentiality concerns.

Nearly all of these sites contained only black abalone shell pile features with no other historical artifacts, save small amounts of lumber planks. In addition to CA-SRI-40, -95, -238/H, -241/H, -353, -471, -559, and -91-75, identified by Morris (1994), my research identified two historical abalone fishing sites with artifacts or features. The first, CA-SRI-799H, contained no black abalone shell pile but rather a large collection of historical artifacts scattered across a large, deflated, windswept point, Cluster Point, on the southwestern coast. This site is located in the same general area as CA-SRI-95, and they may be part of the same locality. Morris described CA-SRI-95 as being completely eroded away, however, making it impossible to evaluate whether the two are from the same site complex. Some of the historical debris is likely from the wreckage of a nineteenth-century wooden vessel, the *Dora Bluhm*, which capsized offshore in the area in May 1910. Our surface reconnaissance identified 24 brown-glazed stoneware food storage jar fragments and three celadon bowl sherds. Morris's earlier work at CA-SRI-95 resulted in the collection of 28 brown stoneware food storage jar fragments. Additionally, Morris collected two brown stoneware jar fragments from CA-SRI-96, adjacent to CA-SRI-95, all of which can probably be safely grouped into a single site assemblage. These are certainly the remnants of a Chinese camp, as are likely the dozens of glass and metal fragments found associated with the Chinese ceramics. A metal grommet, perhaps from a tent, was also identified and lends additional evidence for a Chinese camp. The second site, CA-SRI-804H, is located at Brockway Point and contains two small black abalone shell piles and a distinctive Chinese, hairpin-shaped hearth feature, with historical glass and metal fragments and charcoal embedded in the matrix eroding from the hearth.

In order to supplement the survey results, our research included the first subsurface excavations ever conducted at Santa Rosa Island historical abalone fishing sites. This work was designed, in part, to sample black abalone shell pile constituents and to recover diagnostic artifacts (Figure 5.11), similar to research agendas on San Miguel Island. A 30-liter bulk sample at the large black abalone shell feature at CA-SRI-109, at the tip of Carrington Point, and a 1.0-×-1.0 m excavation unit at the large abalone cluster associated with the *Goldenhorn* wreck (CA-SRI-241/H) produced no artifacts and a shellfish assemblage dominated (~99 percent by weight) by black abalone shell.

FIGURE 5.11. The first two historical abalone fishing camps ever sampled by archaeologists on Santa Rosa Island: (*A*) the author (*left*) and San Diego State University graduate students Linda Bentz (*middle*) and Breana Campbell (*right*) at CA-SRI-109, preparing to map and sample the linear black abalone shell feature exposed along the sea cliff; (*B*) the completed 1.0-×-1.0-m excavation unit at CA-SRI-241/H. Note the large black abalone shells embedded in historical dune sands in the unit side walls. (Composite and images by author)

Santa Cruz Island

Santa Cruz Island is the largest of the Northern Channel Islands, but we know the least about the archaeological record of Chinese abalone fishing there. A number of historical documents, however, suggest that Chinese fishermen camped at Forney's Cove, Prisoners Harbor, Scorpion Anchorage, China Harbor, and a variety of other anchorages along the vast coastline (see Caire 1993:139–140). In addition, historical letters point to early-twentieth-century contracts between island ranching families and Chinese fishermen.

In 1869, one of the early Santa Cruz Island owners, William Barron, sold his interest to ten investors from San Francisco. One of them, Justinian Caire, was a French immigrant who made his fortune selling hardware to gold miners (Caire 1993). By the 1880s, Caire was the sole owner of the Santa Cruz Island Company and ran a successful livestock and ranching industry on the island. He also developed a diversified agricultural operation, which included wine, fruit, and nut harvests (Caire 1993). Always the keen businessman, Caire contracted out the abalone harvest to Chinese fishermen and negotiated a healthy cut of the profits. Historical documents record these transitions in "a letter in existence written to the Caire Co. in 1908, by Ah Poy of Santa Barbara, ask[ing] to renew his old contract at Santa Cruz Island, in the same terms as previously, 'one hundred fifty pounds of every one thousand pounds that are gathered both of meats and shells'" (Caire 1993:140).

Unfortunately, no archaeological record of Chinese occupations has been identified at Prisoners Harbor or Scorpion Anchorage, where historical accounts leave very little doubt that Chinese camps were located (Figure 5.12). Both of these harbors were major centers of transport and ranching operations, and by the mid-nineteenth century contained adobe buildings and ranch headquarters. The expansion of these areas through the nineteenth and into the twentieth centuries probably destroyed all evidence of Chinese campsites and local fishing operations. By 1885, for example, Scorpion Anchorage was described as a "busy and well-equipped colony, which grew to include storehouses, a bake oven, blacksmith shop, wells, a dairy, and several barns" (Livingstone 2006:466).

In the summer and fall of 2013, I spearheaded the first systematic effort to identify and record Chinese fishing sites on Santa Cruz Island;

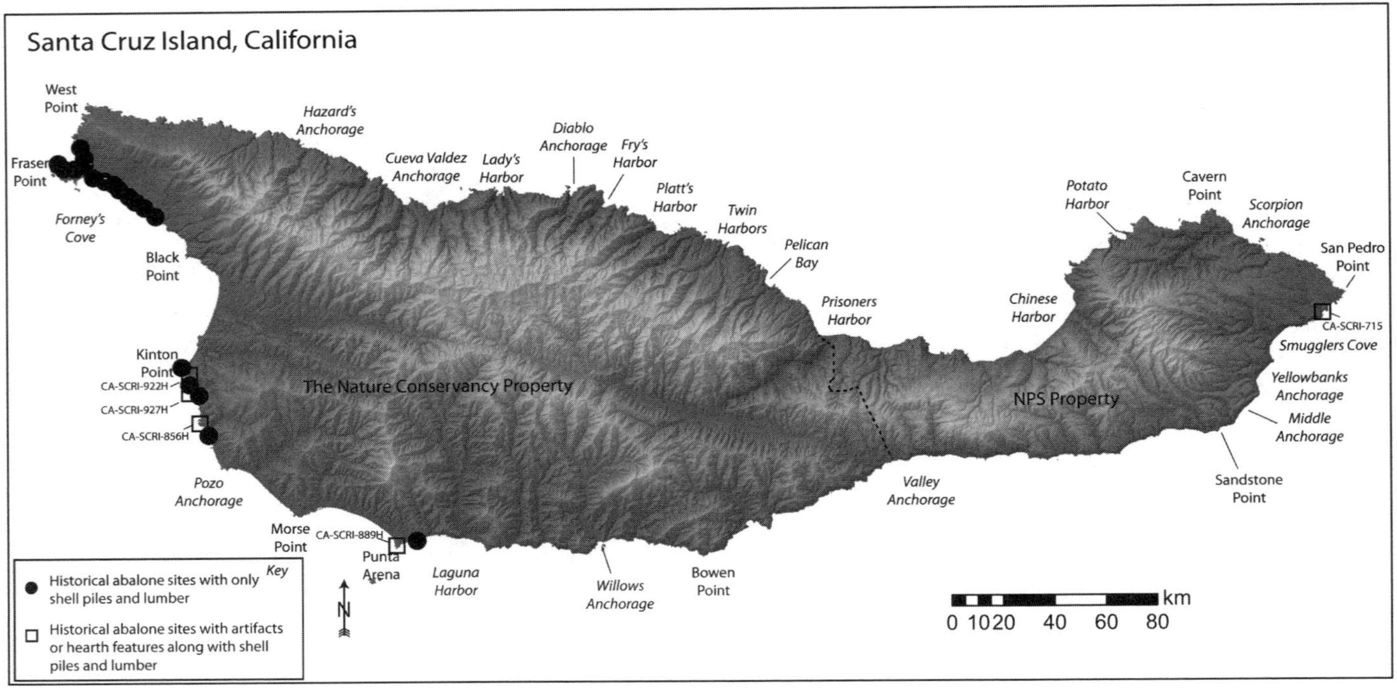

FIGURE 5.12. Map of the 24 historical abalone fishing sites identified on Santa Cruz Island. Five of the sites contain diverse or distinctive historical features and/or artifacts. *Note:* Site locations are approximate due to National Park Service confidentiality concerns.

at the same time, other archaeologists began recording a number of historical Chinese abalone fishing sites as part of coastline surveys of the island's west end. Thus far, small portions of the island coastline have been surveyed, but the vast majority remains uninvestigated. In many ways, archaeological research on Santa Cruz Island has lagged behind efforts on the other Northern Channel Islands. Primarily because of the island's massive size (249 km^2, ~110 km of coastline), relatively little survey work, proportionally, for either prehistoric or historical sites has been completed when compared with the other Northern Channel Islands. In addition, access to island coastlines by land is often very difficult due to high, steep sea cliffs and rugged interior mountain ranges. Logistical and permitting challenges also exist with the east end of the island, owned and administered by Channel Islands National Park, and the west end, by the Nature Conservancy. In a relatively short time, however, a number of historical Chinese fishing sites have been identified along various portions of the island. On the west end, coastlines from West Point to Punta Arena have been surveyed; and on the east end, coastlines from Potato Harbor to Cavern Point, from San Pedro Point to Middle Anchorage, and Scorpion Anchorage have all been surveyed.

In total, 24 historical abalone fishing sites have been documented on Santa Cruz, of which the majority cluster on the northwestern shore of the island (n = 14, 58 percent) and near Kinton Point (n = 7, 29 percent). Despite extensive erosional exposures and surface reconnaissance, the Forney's Cove site cluster has produced only black abalone shell features, with no associated historical artifacts.

The majority of historical abalone sites on Santa Cruz likely represent small Chinese campsites where black abalone were collected for brief periods. One of the five exceptions is CA-SCRI-856H. This historical abalone camp is positioned along the southwestern coast and located on a prominent marine terrace above a small canyon. The locality includes three black abalone shell clusters and what has been interpreted as a 4-×-4-m platform feature with at least 10 upright stakes flanked by more than 35 rock cobbles. Initial reconnaissance and mapping of this site suggest that it may have been a longer-term Chinese residence, but the site has undergone extensive coastal erosion and no other historical artifacts or features such as hearths have been identified.

The largest historical abalone fishing site identified on Santa Cruz is located along the southwestern corner of the island at Punta Arena.

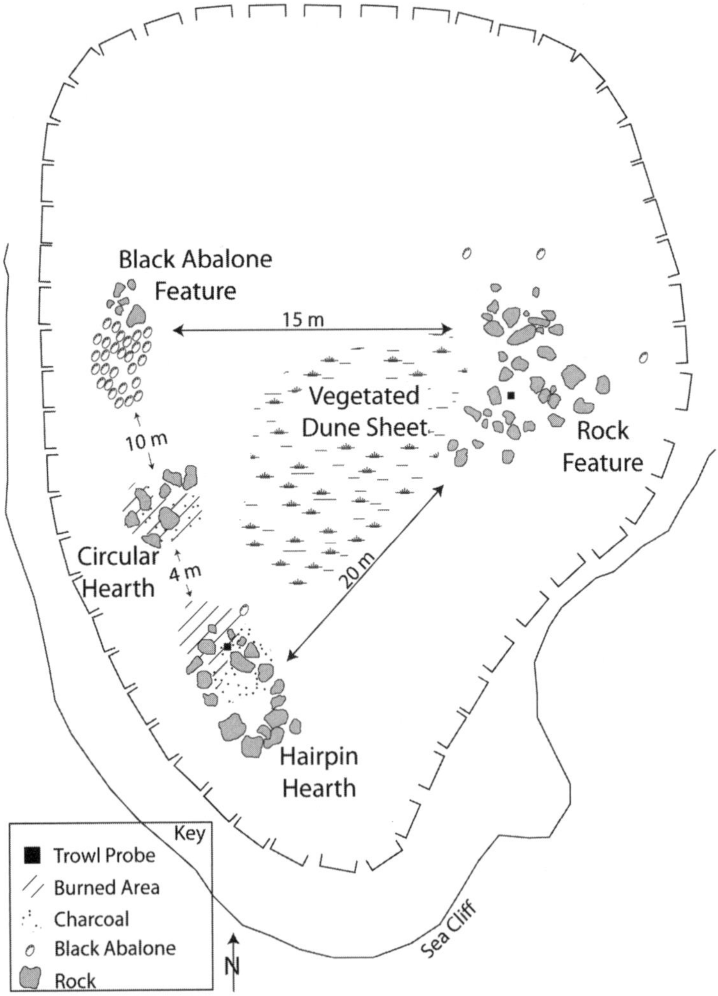

FIGURE 5.13. Sketch map of CA-SCRI-889H, which contains two hearths, a rock feature, and a black abalone shell feature. (Not to scale.)

Exposed on the southern tip of a rocky point is a relatively large Chinese fishing camp, CA-SCRI-889H. The site, although heavily eroded, contains a black abalone shell feature, one hairpin-shaped hearth, one roughly circular hearth, and a large rock feature (Figure 5.13). The abalone shell feature is positioned within 15 m of the two hearths, and their interiors, which contain burned soils and abundant charcoal, were likely used to boil abalone meat. The rock feature's function or functions are impossible to definitively determine, since it is heavily eroded. Some of the rocks associated with this feature are burned, as is the subsurface soil (albeit lightly) within the feature, suggesting that it may have functioned as a third hearth or, alternatively, as the base for large drying racks or a windbreak.

Two other historical Chinese abalone fishing sites have produced evidence of longer-term occupation in the Kinton Point area. CA-SCRI-922H contains only a small remnant scatter of black abalone shells, but surface collections near the site locality produced a bone abalone pry bar and a single brown-glazed stoneware jar fragment. Not far from CA-SCRI-922H, a large historical abalone site, CA-SCRI-927H, contains several black abalone shell scatters and a hearth feature that consists of large angular rocks, burned soil, wood charcoal, and metal fragments, indicating that abalone meat was likely boiled and processed there.

Despite more extensive and systematic survey coverage of the island's east end, not one historical black abalone shell feature was located and only a single Chinese abalone fishing site was discovered. While surveying the coastline at the mouth of a large drainage north of Smugglers Cove, we discovered six small fragments of brown-glazed stoneware on the gully floor at CA-SCRI-715. CA-SCRI-715 was recorded in 2002 as both a prehistoric and historical site containing a number of cultural and natural features. The site consists of two interconnected rock shelters, the lower and larger of which retains evidence of past occupation. Below the mouth of the shelter, in the toss zone, shell, flaked stone, bone, and ground stone are visible in the lower 50 cm of the gully wall exposure. In the upper 10 to 20 cm, bottle glass fragments, rusted tin cans, wire, and other metal debris are visible, along with associated scattered fragments on the eroded slope and gully floor surface. The bottle glass likely dates from the 1920s to 1930s, well after Chinese activities on the island. A historical rock wall, built from water-rolled cobbles and angular rock fall, extends from the mouth of the rock shelter to the edge of the gully drainage. Several early-to-mid-twentieth-century beer cans, amorphous metal fragments, and early-to-mid-twentieth-century bottle glass fragments were found embedded in the rock wall. Inside the rock shelter, an additional small rock wall was constructed, likely as a windbreak, during the twentieth century.

The only evidence of a Chinese fishing component at the site came from the recovery of six small brown-glazed stoneware ceramic sherds, each recovered from a 10-×-10-m area along the gully floor below the rock shelter entrance. During the wet season and times of heavy rainfall, this drainage directs freshwater from the uplands to the ocean, and any artifacts on the gully floor are subject to natural transport and other taphonomic processes. Each ceramic fragment is no bigger than a house key, likely because of fragmentation during heavy rains. To date, these are the only Chinese ceramics ever recovered from a Santa Cruz Island site. Without a black abalone shell feature or other additional evidence, however, it is impossible to determine the primary context of these artifacts and whether Chinese fishermen were using the same rock shelter as the Chumash and later historical fishermen and campers, or whether they occupied some other location upstream or at the gully mouth.

Chinese Abalone Foraging Strategies on the Northern Channel Islands

Based on archaeological field surveys, excavations, and historical document searches, a pattern of Chinese abalone fishing and island exploitation has started to emerge. The vast majority of historical Chinese abalone fishing sites on the Northern Channel Islands contain only piles of black abalone shells, with no associated artifacts, hearths, or other features. Some of these sites, however, contain scattered planks of lumber associated with abalone piles, likely the remnants of abalone meat drying racks. Subsurface excavations of black abalone features at these site types suggest that they were highly specialized, commercial fishing localities focused on black abalone harvesting (Braje and Erlandson 2007). Despite the local availability of a diverse array of shellfish (e.g., California mussels, sea urchins, limpets, gooseneck barnacles), fish (e.g., California sheephead, rockfish, surf perch), and seabirds (e.g., gulls, pelicans, ducks) in historical times, there is no evidence that Chinese fishermen exploited these resources for either commercial or subsistence purposes. These site types were likely highly specialized logistical fishing locations, which Chinese fishermen occupied for brief periods to collect and initially process intertidal black abalone. Once a local rocky intertidal was depleted, these nonresidential logistical processing sites were moved down the coast.

A small number of sites produced a more diverse array of historical artifacts and features (see Chapter 6). Most of these locations contain black abalone shells and lumber, along with metal fragments, bottle glass, ceramic sherds, and hearths. The restriction of features and artifacts to these locations, often located adjacent to protected sandy beaches or harbors, suggests that they acted as long-term basecamps from which abalone fishermen logistically harvested the intertidal. Basecamps were ideally located adjacent to sandy beaches or protected harbors and served as landing points for Chinese junks or Euro-American vessels and launching points for abalone collecting and processing forays around island shorelines. Historical documents indicate that Chinese fishermen were dropped off on the islands by junks (or schooners) at protected harbor locations. Boiling kettles, sampans, foodstuffs packaged in brown-glazed stoneware vessels, tents, and personal items were unloaded for a three-to-four-month stay on the islands. Chinese fishermen would have maintained a presence at these basecamps, where additional supplies could be dropped off and where vessels retrieved the fishermen for their journey back to the mainland. Since intertidal black abalone communities could be quickly and efficiently fished out from a local watershed, the fishermen needed to continually travel down the coastline, farther and farther afield from their basecamps, to locate new stocks.

While historical documents clearly describe that Chinese fishermen were boiling and drying abalone meat on the islands before transport back to the mainland, surprisingly few historical abalone processing sites contain evidence of hearth features. Likely, abalone collected from logistical camps were being dried (at least partially) near the point of collection but then transported back to basecamps for intensive boiling and drying. Boiling required large kettles that were difficult to transport, so this foraging pattern probably grew out of practical necessity. Along coastlines with steep and treacherous sea cliffs, such as the northeast and eastern shores of Santa Cruz, much of the northern shore of San Miguel, and northwestern Santa Rosa, logistical fishing camps and abalone transport to central locations for boiling were likely accomplished with sampans. In areas lacking gentle coastal plains or easily accessible locations out of the supratidal zone where initial processing could be accomplished, very little archaeological record would remain. This could explain the lack of historical abalone fishing sites on eastern Santa Cruz, northern San Miguel, and portions of northwestern Santa Rosa. The abundance of logistical processing sites on southern San Miguel Island and eastern Santa Rosa and Santa Cruz Islands probably is related to the presence of gentle coastal plains in these areas, where initial abalone processing could be accomplished along sea cliffs and rocky points. Access to these locations was possible both by sea, in sampans, and very easily and quickly by foot.

These physiographic factors would have been important in shaping Chinese fishing patterns and the archaeological record. It is also important to remember, however, that the location of logistical processing camps would have been dependent on the presence of productive intertidal habitats. The vast sandy beaches along northeastern Santa Rosa, for example, were probably not prime abalone fishing habitat. In addition, we cannot discount other social or environmental factors that may have helped create this record. On San Miguel Island, for example, the majority of historical Chinese fishing sites are located on the southern shore. It may be that Chinese fishing activities were concentrated along this portion of the island to avoid ranching operations centered on Cuyler Harbor and the north coast. If this was the case, it was not a universal pattern, since mutually beneficial contracts between some island ranch owners and Chinese fishermen were established by the turn of the twentieth century.

In the next chapter, we explore the two largest and most diverse Chinese abalone fishing basecamps yet identified on the Northern Channel Islands. Basecamps on south-central Santa Rosa Island, CA-SRI-471, and western San Miguel Island, CA-SMI-614/H, offer unique insights into the lives of nineteenth-century abalone fishermen on the Northern Channel Islands. Recovery and analysis of material culture and features from these sites help contextualize who these fishermen were, their strong connections to their recently departed homelands, and their island activities. By interpreting these data, we can better understand what life was like for a nineteenth-century Chinese abalone fisherman on the Northern Channel Islands. Archaeological research at CA-SMI-614/H also helps paint a vivid picture of the changing economic and social landscape for the Chinese as they were pushed out of the abalone fishing industry beginning in the late nineteenth century.

CHAPTER 6

Nineteenth-Century Abalone Fishing on California's Channel Islands

The Archaeology of Johnsons Lee and Point Bennett

Early in the morning of December 7, 1941, the Imperial Japanese Navy launched a surprise attack against the U.S. naval base at Pearl Harbor, Hawaii. Before this deadly assault, the United States had activated a number of military defenses along the Pacific Coast, but the Pearl Harbor attack spurred California to rally its coastal defenses and spawned a network of lookout stations and aircraft warning installations from San Diego to Canada (Livingston 2006:264). Santa Barbara was on especially high alert. Only three months after Pearl Harbor, a Japanese submarine surfaced about one mile off the coast and launched 25 cannon shells at the Ellwood Oil Field oil-storage tanks in Tecolote Canyon (Erlandson, Rick, and Vellanoweth 2008:113; Tompkins 1966:305). Most never detonated, and those that did caused only minor damage, but the event remains the only foreign naval attack against the continental United States since 1812. In the aftermath, the U.S. Navy established coastal lookouts on all of the Northern Channel Islands and a radar post on Santa Rosa Island.

Johnsons Lee, Santa Rosa Island

Although the military no longer maintains facilities on any of the Northern Channel Islands, its historical activities left a significant footprint on island landscapes. One of the largest military bases was constructed in the early 1950s at Johnsons Lee on the central southern shore of Santa Rosa Island. This extensive air force base included five two-story barracks for airmen, a dining hall, training and recreation buildings, offices, warehouses, and a dispensary, all built near the island shore, along with a solid concrete pier (Livingston 2006:274–275). A paved five-mile-long roadway 14 feet wide and complete with metal guardrails, a drainage system, and dozens of culverts connected the Johnsons Lee facilities with mountaintop transmitters on the summit of Vail Peak to the north (Livingston 2006:363). By early 1952, the base facilities were complete and operated as a critical component in detecting, identifying, and tracking air traffic in southern California for its defense against enemy threats. The Johnsons Lee facilities became a critical island hub, providing basic necessities and recreational activities for military and other island visitors, even scientists working on the island. For example, one of the earliest and most influential archaeologists to conduct research on Santa Rosa Island, Phil C. Orr, regularly drank at the Air Force Officers Club by traveling jeep tracks and the navy road from his basecamp on the northwest coast.

After little more than a decade, the air force deactivated operations on Santa Rosa and closed its island facilities, citing exorbitant operation costs and logistical challenges. More than 40 buildings and structures were abandoned to the island landowners, the Vail and Vickers families, who repurposed construction materials and structures for their island ranching operations. For 30 years, the buildings and associated materials left behind were picked over and slowly deteriorated until the National Park Service acquired the island in the mid-1980s and began a nearly 10-year effort to remove the buildings and pier and revegetate the area (Livingston 2006:284).

The remnants of this military history still loom large. The trek from the modern pier at Bechers Bay on the north coast to Johnsons Lee takes visitors past deserted upland navy installations and down the now dilapidated navy road, following a circuitous route through a large canyon. Most of the road remains intact, but the asphalt surface is cracking, potholed, and eroding away along its edges; the drainage system is failing; and the guardrails have been repurposed for cattle corrals. For the uninitiated, it can be a white-knuckle drive down the steep canyon. The road terminates directly north of Johnsons Lee, where the once sprawling air force facility stood. Only one building remains, a more than 2,000-square-foot auto maintenance shop, now used by the National Park Service for limited storage and remote operations (Livingston 2006:285). Scattered debris can still be found along the sea cliff and supratidal, but the once massive concrete pier has been completely removed, leaving historical documents and pictures as the best indicators of the tremendous scale of military activities in the area.

Given the number of buildings and military personnel who once lived and worked at Johnsons Lee, it is surprising that any archaeological sites remain. Less than 800 m to the west, however, past a small drainage, sits the heavily eroded remnants of one of the most significant historical Chinese abalone fishing sites on the Northern Channel Islands (CA-SRI-471). Unfortunately, little remains of the once sizable Chinese basecamp, not because of military activities but due to the ravages of historical overgrazing and erosion.

CA-SRI-471 is positioned about 100 to 150 m inland from the coast, eroding from the center of a large arroyo complex where the occupants established their camp to seek shelter from the wind. The site surface is devoid of vegetation, save several islands of soil and sediment with a sparse vegetation cover of low grasses and shrubs. The arroyo probably began forming during the historical ranching period after the area was devegetated by domesticated animal grazing. Steady wind and water erosion over the last 150 years has created a badlands terrain and melted away the historical living surface of the Chinese inhabitants. Lost or discarded artifacts, abandoned features such as hearths or windbreaks, and other remnants of past human activities have been washed steadily downslope and transported, in some cases, very long distances from their likely points of origin. In most cases, these natural transformation processes ruin the integrity of archaeological sites and result in limited research potential.

The main site area consists of a small rock pile and concentration of historical debris covering an area of about 30 m^2. Erosion has been so extensive, however, that materials are spread downslope over an area of at least 400 m^2. While immensely destructive, these processes have contributed to what makes CA-SRI-471 so remarkable. Erosion has created exceptional visibility and allowed for the surface collection of some Chinese artifact types that have never before been recovered on the Northern Channel Islands.

The most abundant artifacts recovered from the site deposits are 30 brown-glazed stoneware fragments, all from medium-sized storage jars. Detailed laboratory analysis allowed for the refitting of the majority of

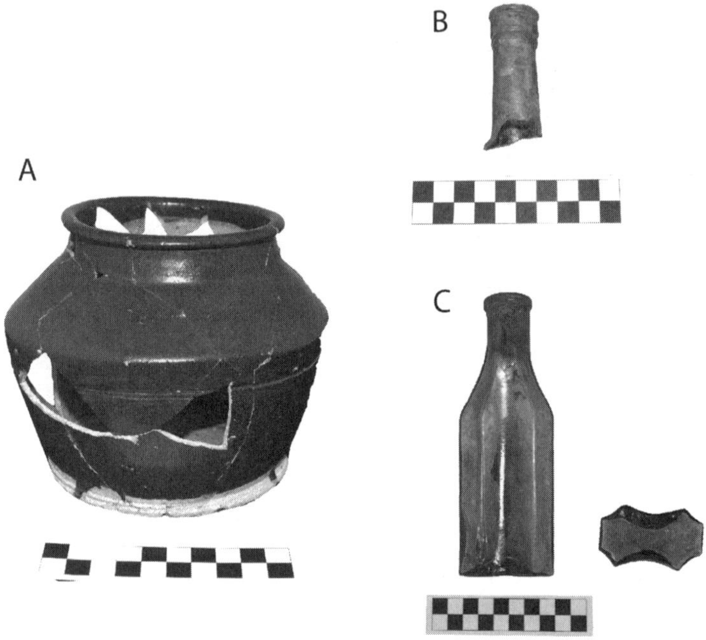

FIGURE 6.1. A sample of the historical artifacts recovered from CA-SRI-471: (*A*) refitted medium-sized, large-mouthed storage jar; (*B*) neck and top of a pepper or sauce bottle; (*C*) whole "spice" bottle.

these to re-create a single wide-mouth jar. This jar, once used to transport preserved foodstuffs to the island, is the most complete example of a Chinese vessel from the Northern Channel Islands (Figure 6.1A). During its life, it likely contained tofu, beans, pickled vegetables, or some other commonly consumed food transported by Chinese abalone fishermen to southern Santa Rosa Island.

From a large and highly fragmented collection of broken glass, amorphous metal, bullet casings, a hand-forged pre-1880s nail, and eroded storage battery parts, six artifacts stand out. Two of these are glass artifacts from bottles that once contained sauces or condiments. Such bottles are common in nineteenth-century archaeological sites, since food during this time was often bland and of dubious quality. A wide range of sauces and spices were marketed to enhance the flavor of foods or to cover up distasteful ones.

It seems that the Chinese fishermen at Johnsons Lee were no strangers to the benefits of these sauces and spices. One of the glass artifacts recovered from the site deposits is the fragmented neck and top of a light green pepper or sauce bottle. These are one of the most commonly found "food" bottles in archaeological sites in North America during the nineteenth and twentieth centuries (Figure 6.1B). The finish of the top and seam along the body of the bottle are distinctive in this bottle type, and the CA-SRI-471 example likely dates to the early 1870s to the mid-1880s.

Remarkably, surface reconnaissance recovered an intact aqua blue "spice" bottle with the remnant cork stopper still adhering in place (Figure 6.1C). Spice bottles are very distinctive, and although they came in a variety of sizes from about 2.5 to 14 ounces, all are eight-sided (viewed from the base) with wide, concave front and back panels and flat, narrow side panels joined by wide, concavely beveled corners (Fike 1987). These bottles are commonly found in mid- to late-nineteenth-century archaeological sites and typically contained spices such as thyme, cloves, cinnamon, or marjoram or condiments such as ketchup or mustard (Zumwalt 1980). Narrowing down the exact contents is impossible, however, since no labeling or embossing remains. Analysis of the manufacturing process used to produce the container suggests that it dates between the 1870s and the early 1880s.

Given that both of these glass bottles date the site deposits between the early 1870s and early 1880s, it may be that the Johnsons Lee historical basecamp was occupied during the heart of the Chinese abalone fishery in southern California. The earliest newspaper account of Chinese abalone fishing on the Northern Channel Islands comes from an 1856 account in the *Sacramento Daily Union*: "At 7:30 PM, September 28th, below Monterey, during a thick fog, [a steamer] ran down the schooner *Frances Skiddy*, of 17½ tons. The Captain, Hugh Robertson, was drowned. The schooner was bound to Santa Cruz Island with a cargo of provisions for the Chinese fishery, and had on board four Chinese passengers, who were all saved." If not for this tragic accident, Chinese abalone fishing would have gone another year without historical documentation until several media outlets (e.g., the *Hutchings's Illustrated California Magazine* and the *Marysville Daily Herald*) described Chinese fishing activities on the California Channel Islands in 1857. The *Sacramento Daily Union* account suggests that Chinese abalone fishing was established by at least 1856, and the first island fishing camps probably coincided with the earliest mainland camps in 1853 (see Chapter 4).

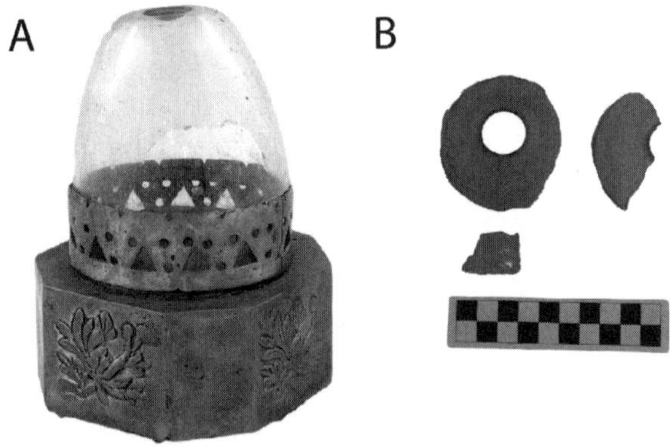

FIGURE 6.2. (*A*) Nineteenth-century opium lamp (Open access image from the University of British Columbia Library, http://digitalcollections.library.ubc.ca/cdm/ref/collection/chung/id/8016). (*B*) Three glass fragments from at least two opium lamps recovered from CA-SRI-471.

Scattered among the historical debris were three glass fragments from at least two different opium lamps, identified by laboratory analysis (Figure 6.2). By the mid-nineteenth century, opium smoking was widespread among Chinese immigrants throughout the American West. Opium was not grown in the United States but imported in raw and cooked form and heavily taxed at a rate of up to 182 percent of its value (Courtwright 2001:16). For many archaeologists, historians, and members of the general public, opium use among early overseas Chinese communities has been a point of fascination. The topic conjures images of depraved Chinese addicts wasting away in sordid opium dens. The reality is that opiate use in the United States was pervasive and much more extensive than among only one ethnic community. During the Civil War, for example, 10 million opium pills and over 2,841,000 ounces of opium powders and tinctures were issued to Union forces (Courtwright 2001:54). In the mid-nineteenth century, 24.5 percent of prescriptions written by New Orleans doctors contained opiates, which were commonly added to patent medicines throughout the country (Courtwright 2001). Opiate use was widespread in the nineteenth century, but opium smoking occurred almost exclusively in Chinese communities and served as a social enterprise in a range of communal contexts (Stirling 1913).

The opium glass fragments from Johnsons Lee came from lamps, examples of which are commonly found at historical Chinese archaeology

FIGURE 6.3. Historical photograph of a Chinese man smoking opium. Note the glass top of the opium lamp and the opium pipe positioned upside-down over the lamp. (Open access image from Thomson 1898:159)

sites throughout the American West. In order to be smoked, opium had to be warmed over a heat source and, once malleable, placed into a pipe with a specially designed pin. The pipe was then positioned upside-down over an opium lamp so the vaporized flumes could be inhaled (Figure 6.3; see Courtwright 2001 and Stirling 1913). Since no other opium implements, such as pipes, trays, or containers, were identified at Johnsons Lee, these three fragments of opium lamp glass remain the only examples of opiate consumption ever recovered at a historical Chinese abalone fishing site on the Northern Channel Islands.

The final unique discovery at CA-SRI-471 and Johnsons Lee is the highly eroded remnants of a shoe sole. The vestiges of this leather-soled shoe were found partly buried below a driftwood log several meters to the west of the central rock cluster. The preservation of this artifact is remarkable, especially in an open-air site where the exposure of perishable material to the elements tends to expedite biodegradation. As the area was devegetated by overgrazing and as water run-off cut deep channels across the site surface, this discarded shoe became partly buried below a driftwood log, likely brought to the site by Chinese fishermen from a local beach for firewood or as a makeshift seat. The dry Channel Islands climate and the protection offered by partial burial and the log allowed for unusual preservation conditions. Its heavily eroded state still makes definitive identification impossible, but the ~25-cm-long leather sole appears to be from a hand-stitched, slipper-type shoe, which was likely attached to a cloth upper. Given its association with other clearly Chinese artifacts and the lack of later historical material (e.g., military period artifacts), the shoe sole is likely the discarded remnant of a common nineteenth-century Chinese shoe.

Beyond these outstanding finds, what the Chinese basecamp at Johnsons Lee lacks is the preservation of historical features, such as hearths or abalone piles. This is almost certainly related to the massive scale of erosion that has transformed the historical living surface and scattered archaeological materials hundreds of meters downslope. Fortunately, at the far west end of San Miguel Island rests a spectacular Chinese abalone fishing basecamp that can help fill this gap and offers exciting new insights into the lifeways of island Chinese abalone fishermen.

Point Bennett, Western San Miguel Island

Point Bennett, positioned at the far western extent of San Miguel Island, is a magical place for scientists and tourists alike. Modern island visitors often moor at the largest, most protected anchorage on San Miguel Island, Cuyler Harbor, along the north-central coast. The adventurous brave a 14-mile round-trip trek to stand at the edge of a plateau 100 feet above Point Bennett. From here, you can gaze out onto the vast Pacific over a forbidding landscape of windswept beaches, rocky headlands, and constantly shifting sand dunes. The beaches and offshore kelp forests teem with the greatest variety of pinnipeds found anywhere in the world: northern fur seals (*Callorhinus ursinus*), northern elephant seals (*Mirounga angustirostris*), harbor seals (*Phoca vitulina*), California sea lions (*Zalophus californianus*), and Steller sea lions (*Eumetopias jubatus*). If you are lucky, you even might see a Guadalupe fur seal (*Arctocephalus townsendi*), an occasional visitor today but, according to archaeological data, abundant in the past (Braje and Rick 2011; Rick et al. 2009). Gazing on one of the largest pinniped rookeries in the world with a pair of binoculars, it's hard not to lose track of time watching these animals fight, sun themselves, care for their young, and laze on sandy beaches.

For hikers, this is the end of the line. Access to Point Bennett is tightly restricted, and even archaeologists and other scientists must have explicit permission, an approved research project, and a sea mammal biologist escort to continue down the steep trail. Those few archaeologists who have conducted fieldwork at Point Bennett have uncovered a rich human history that extends back at least 10,000 years. Fourteen sites have been recorded and radiocarbon-dated to between 10,500 and 7,500 years old, and scores of deeply stratified shell middens dating back 7,500 years are embedded in a series of northwest-southeast-trending sand dunes (see Erlandson et al. 2005; Kennett 2005; Walker et al. 2002).

One of the largest and most impressive sites in the Point Bennett area is CA-SMI-602, a shallow, stratified shell midden and house pit village located in the midst of the modern rookery immediately north of Adams Cove. This village was first established by the Chumash Indians, the Native American occupants of the Northern Channel Islands and adjacent mainland, about 500 years ago and was an important fishing, shellfishing, and marine mammal hunting community until AD 1660 (Walker et al. 2002:629). The remnants of at least seven large house pit depressions are still visible and speak to the dramatic changes that have taken place along Point Bennett in just the last several hundred years.

Today a sea mammal sanctuary, the Point Bennett landscape was dominated by people until the mid-seventeenth century. The Chumash and their ancestors lived, worked, and played along the beaches and sand dunes for over 10,000 years. Their populations grew and they gradually transitioned from seasonal visitors to permanent residents (see Rick et al. 2005). Their presence would have restricted sea mammals to offshore rocks and islands, where the threat of human hunters was less immediate. It was not until after the abandonment of Point Bennett by the Chumash that sea mammals began to inhabit the beaches. Their numbers, because of rampant overhunting by fur traders and sealers, were just a tiny fraction of what we see today.

Evidence for the next substantial human occupation comes 200 years later, southeast of the last vestiges of Chumash livelihood. Located behind the beach at Adams Cove is the scattered archaeological evidence of a

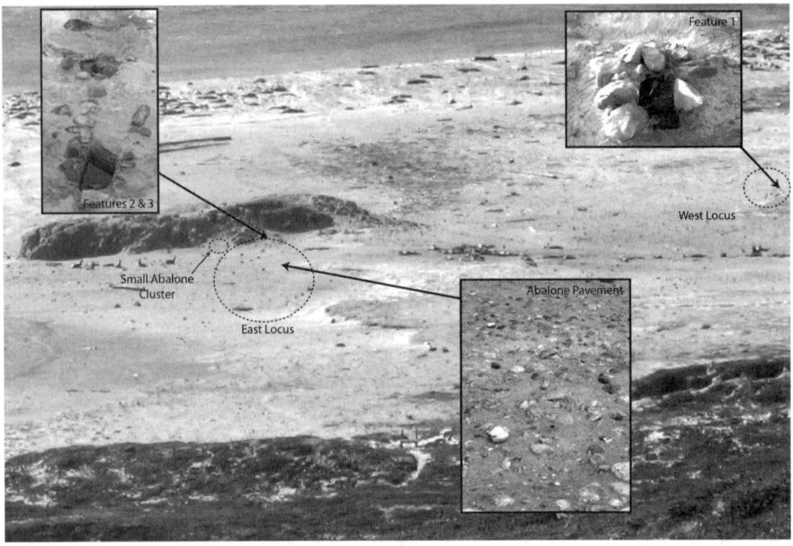

FIGURE 6.4. Overview photograph of the East and West Loci at CA-SMI-614/H and inset photographs of Features 1, 2, and 3 and the abalone pavement. Note the sea mammals (5–6 feet long) on the beaches for scale.

large historical hunting and abalone processing site, CA-SMI-614/H, which contains multiple occupation components scattered over an extensive area. Storm surges, coastal erosion, and pinniped activities have disturbed the site and mixed historical artifacts with modern flotsam and jetsam and ancient Native American material culture (Braje and Erlandson 2008; Braje, Erlandson, and Rick 2007; Braje et al. 2014).

Despite these taphonomic obstacles, two clear historical loci have been identified. The East Locus is situated in the eastern lee of a large rock outcrop containing two hairpin rock hearth features (Features 2 and 3), a pavement of mostly whole red and black abalone shells covering at least 108 m², a smaller pavement of whole black abalone shells covering 16 m², and a wide, thin scatter of prehistoric and historical artifacts covering 3,750 m² (Figure 6.4). The West Locus, smaller and much more concentrated, is located about 200 m to the northwest, on the southern margin of CA-SMI-602. The locus contains a roughly circular rock hearth feature (Feature 1) and a light scatter of mostly historical artifacts.

Although the specific function or functions of these rock features are impossible to determine, excavation of their interiors revealed underlying burnt and mottled soil with charcoal and ferrous oxide stains (see Braje et al. 2014), suggesting that all operated as hearths. Just outside the eastern opening of Feature 3, excavations exposed a 12-inch-diameter burnt circle of oily, charcoal-rich matrix. The interior matrix of hearth Feature 1 contained abundant charcoal, burnt soil, a variety of nondiagnostic metal fragments, black abalone shell fragments, a whole black abalone shell, and a large piece of flat sheet metal. Excavations and mapping also revealed a rock wall alignment extending to the northwest from the main, circular hearth feature. Several metal fragments, a carved wooden stake, a whale bone fragment, and a cluster of articulated sheep foot bones were discovered embedded within the rock alignment.

Scattered throughout the site, over 350 historical artifacts were recovered from the East and West Loci. Analysis of pottery sherd associations from these artifact assemblages demonstrated that fragments from the same vessel were spread over virtually the entire East Locus, likely the result of storm surges, pinniped activities, and other modern disturbances. This makes fine-grained analysis of artifact distributions to interpret activity areas impossible. There are some distinct differences in the ages and composition of historical artifacts, however, between the East and West Loci.

Ceramic artifacts were the most abundant artifact type recovered from both the East and West Loci. At the East Locus, over 50 Chinese brown stoneware pottery fragments were recovered from a variety of vessel types, including foodstuff vessels of various sizes and shapes, a liquor bottle, and a ginger jar, of a type often packed with ginger and other commodities such as salt.

The East Locus produced not only the vessels that Chinese fishermen used to transport foodstuffs to the island but also the distinctive tableware vessels used during meals. Like the brown-glazed stoneware vessels, these tableware "rice" bowls were all imported from China. Three were decorated with the common bamboo pattern, two with celadon glazes, and one with a double-happiness design.

A wide variety of glass fragments were also recorded from the East Locus, representing 20 individual jars and bottles. These include wine bottles, a soda bottle, an olive-oil bottle, food jars, canning jars, and a demijohn. A few of the bottle fragments retain manufacturer marks that can help identify their contents and when they were produced. A shoulder seal embossed "ELLWOOD COOPER/PURE OLIVE OIL/SANTA BARBARA, CA" from an aqua-glass fragment, for example, identifies a famous promoter of California olive oil and dates between ca. 1880 and 1892 (Carter 2007; Erlandson, Rick, and Vellanoweth 2008:107). Analysis of other bottles and makers' marks suggests an early-to-mid-twentieth-century occupation (see Braje et al. 2014; Toulouse 1971:220).

In addition to ceramics and glass, the East Locus produced a diverse array of personal items, firearms equipment, clothing items, and structural artifacts. These include brass cartridge cases, men's shirt-type buttons (six glass and three abalone shell), a brass/copper suspender clip, and a men's work clothing button (Psota 1999:110). Intermixed with other artifacts were nails and hardware fragments, some of which may have been used in wooden boat construction or as part of a historical structure.

Perhaps the most intriguing and enigmatic artifacts recovered from any historical site on the Northern Channel Islands are the 67 sea mammal teeth artifacts collected from CA-SMI-614/H's East Locus. Aside from four whole teeth (two canine and two caniform incisors [precanines]), all of the teeth have been sawn and modified in an effort to produce ivory rings or pipe stems. Using comparative collections at the National Marine Mammal Laboratory in Seattle, the NOAA marine mammal biologist Robert DeLong and I determined that all the identifiable elements (approximately half the total) were constructed from Steller sea lion teeth. Steller sea lions are a rare visitor to the Channel Islands today but were abundant in the early twentieth century.

Analysis reveals a clear production sequence of ivory rings from the canines (Figure 6.5) and a likely production sequence of ivory pipe stems from the precanines. Whole Steller sea lion canines were collected, their distal and proximal ends sawn off and discarded, and their midsections cut into medial ring blanks a quarter- to half-inch thick. The blanks were then carved and polished into beautifully intricate ivory rings, smoothed and flattened on the inside and slightly beveled on the exterior. This entire production sequence is represented at the East Locus and includes two fragments of broken rings in progress and two completed, but cracked and discarded, rings.

It is less clear how the eight precanine fragments were being used. Most of the debris is not diagnostic, including three medial and two distal fragments and two whole precanines. Based on their smaller size, however, the precanines were probably impractical as ivory rings. The only clue

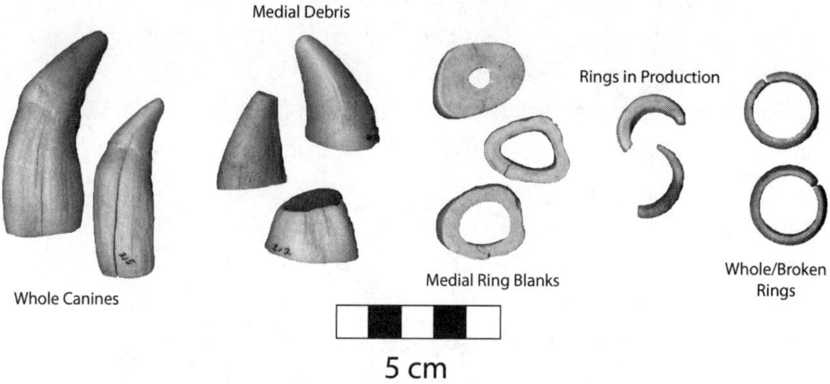

FIGURE 6.5. Sawn Steller sea lion teeth in various stages of production recovered from the East Locus at CA-SMI-614/H.

comes from a single ivory pipe stem in production made from a precanine tooth, suggesting the manufacture of smoking pipe stems. The pipe stem in production is approximately 4 cm long and sawn at both ends. The proximal end has been shaped into a mouthpiece with a constricted neck that has been carved and ground, providing a clear visual separation between the pipe stem body and the mouthpiece. The distal end of the pipe stem was partially drilled adjacent to a natural hole that runs through the tooth. Presumably this was done to widen the opening, but it seems to have resulted in the fracture and discard of the tooth.

The East Locus of CA-SMI-614/H appears to have served as an important Chinese abalone fishing camp during the late nineteenth century. Chinese fishermen employed traditional containers (brown-glazed stoneware) to import foodstuffs that were consumed using familiar Chinese vessels (bamboo-patterned and celadon-glazed rice bowls). Distinctive hairpin-shaped hearths (Features 2 and 3) are similar to Chinese hearths found elsewhere on the Channel Islands (see Berryman 1995:234–256), including at Harris Point on San Miguel and Brockway Point on Santa Rosa (see Chapter 5). These hearth types have also been identified as diagnostically nineteenth-century Chinese on the mainland. Chinese sites in the Mother Lode area of Gold Rush California, for example, have produced hairpin-shaped hearths, depicted in a nineteenth-century lithograph (Costello et al. 1998:Figs. 7a, b) and recorded at a roughly contemporaneous shrimp fishing site near San Francisco (Schultz 1988).

When Chinese fishermen lived and worked at Adams Cove, a large kettle or pot probably rested atop the rock arms of the hairpin hearth and was used to boil abalone meat before it was dried in the sun and wind. The burned circle at the mouth of hearth Feature 3 is likely the remnants of charcoal and debris periodically cleaned out from under the pot or kettle to allow airflow to the hearth. The pavement of black and red abalone shells is all that remains of a once massive abalone pile that probably accumulated in the area. Most of the shells were shipped off the island, to be sold in local markets in Santa Barbara and San Francisco, as well as around the world. The relatively safe harbor and sandy beaches made Adams Cove the perfect location to anchor boats, provision the fishermen, and load their products for transportation to mainland markets.

The Chinese were not the only ones to occupy the East Locus at Adams Cove historically, however. Manufacturer marks on two glass jar bases place their production long after the Chinese abalone fishing industry, one between 1930 and 1950 and one after 1945. Euro-American ceramics provide further evidence of a post-Chinese occupation, likely in the mid- to late twentieth century. Historical diaries of life on San Miguel and the other Channel Islands describe an abundance of camps or cabins for fishermen, smugglers, rumrunners, and eccentric hermits who episodically used the islands for a variety of aboveboard or illicit activities. A historical photograph, for example, was taken in June 1927 at Point Bennett of three Euro-American fishermen standing near a makeshift cabin and drying rack (see Braje 2010). Sea lion "trimmings" (the genitals, gallbladder, and whiskers of bull seals), which were sold to Chinese markets in San Francisco for $2 to $5 a set (Ellis 2003:187), can be seen dangling from the rack. The penis and testicles were powdered and used as a virility aid, the gallbladder was consumed for various medicinal purposes, and the whiskers were employed as toothpicks or opium pipe cleaners. The pelts and meat of these animals had little commercial value, so they were left to rot along the beaches. While the cabin and drying rack have long been destroyed, the jars, gun shell casings, and nails may be the last lingering evidence of these historical hunters.

The presence of ivory rings and pipe stems fashioned from sea lion teeth is the most difficult discovery to explain. Exhaustive literature searches of nineteenth- and twentieth-century mainland Chinatown excavations produced no evidence of similar artifacts, nor did other historical sites of similar age across California and the North Pacific Rim. These may have been trinkets fashioned out of boredom and opportunity, as a way to pass the time on an island where Steller sea lions, especially during the late nineteenth and early twentieth centuries, were uniquely abundant. The finished, but cracked and discarded, rings suggest that as the ivory dried, they were susceptible to breakage, perhaps making this a short-lived and informal industry. Although the ivory artifacts were found scattered among Chinese artifacts and abalone fishing features, it is equally as likely that they were produced by early-twentieth-century Euro-American hunters targeting bull sea lions for their trimmings as by earlier Chinese abalone fishermen.

Though a much smaller assemblage, historical artifacts recovered from the West Locus at CA-SMI-614/H tell a much different story. Only two fragments of Chinese brown-glazed stoneware were recovered from this area. The most distinctive artifacts, however, are two fragments of a medium-sized bowl fashioned in Japan during the Meiji period, ca. 1868 to 1912 (Costello and Maniery 1988:19, 24–25). The sherd fragments are decorated around the exterior with four blue, transfer-printed, round medallions, and the rim's interior and exterior are colored blue through a process of blowing pigment onto the vessel. The exterior of the ring base is decorated with a single blue line. Encircled by a thick blue border, the two identifiable medallion landscape designs are (1) "Man and Mountains," showing a meditating man sitting on an open hillside with mountains in the distance; in the foreground is a rock with leafless tree branches depicting winter; and (2) "House and Rock," depicting a large rock with leafless branches in the foreground and a house in the background to the right (Figure 6.6).

Besides a bullet and brass gun cartridge and a plain brass clothing fastener, the only other artifacts collected from the West Locus were a small set of glass fragments. An opaque glass canning-jar lid fragment likely postdates the late nineteenth century, when such lids surged in popularity (Jones and Sullivan 1985:14, 164–165). Two amethyst jar fragments date between ca. 1903 and 1915, and a whole, amber screw-top bottle manufactured with an automatic bottle machine postdates ca. 1904. Fragments from a rectangular, amethyst-colored medicine bottle were also recovered; one side is convex and once was covered by a label. The reverse side is flat and is embossed with "3iv." The base is marked "Illinois Glass Co.,"

FIGURE 6.6. Photograph of a Japanese hardhat-diving operation at San Miguel Island from a 1913 issue of *Overland Monthly*. *Inset:* (*A*) Japanese bowl recovered from the West Locus at CA-SMI-614/H; (*B*) detail of "Man and Mountains" medallion design; (*C*) detail of "House and Rock" medallion design.

which dates the bottle between 1916 and 1929 (Toulouse 1971:264). Finally, a whole, rectangular aqua-glass machine-oil bottle with cork closure was found during surface collections. The bottle side panels are embossed "THREE IN ONE" on one side and "C.W.COLE CO." on the opposite. 3-in-One machine oil was patented in 1894 by C.W. Cole and used to lubricate, prevent rust, and clean engine equipment. Since Cole incorporated in 1899 and the original cork cap was converted to a screw top in 1910 (Munsey 2006), the West Locus bottle dates between 1899 and 1910.

The West Locus of CA-SMI-614/H helps illustrate the history of commercial abalone fishing in southern California immediately following the Chinese. Based on the small scatter of domestic artifacts and the circular stone hearth, this locality likely dates between 1900 and World War II. During this interval, state-sponsored racism and targeted legislation and taxes drove Chinese fishermen out of the abalone industry (see Chapter 7). The Japanese bowl, the only of its kind ever found on the Northern Channel Islands, evinces Japanese fishermen, who filled the open niche vacated by their Chinese predecessors. An article published in *Overland Monthly* in 1913, titled "Steaks and Pearls from the Abalone," contains a photograph of six Japanese hardhat fishermen diving for abalone off the coast of San Miguel Island (Edholm 1913:383). It may have been this group of fishermen who left the ceramics and used the 3-in-One machine oil to maintain their boat engine or compressor. The circular stone hearth may have been where they huddled for warmth during their time on the island, perhaps even poaching the occasional sheep from island ranchers.

Life as a Chinese Abalone Fisherman on the Northern Channel Islands

Archaeological investigations at Johnsons Lee and Point Bennett help re-create what life might have been like for nineteenth-century Chinese abalone fishermen on the Northern Channel Islands. Stranded at island camps for months at a time, Chinese fishermen combed the intertidal for black abalone with no means to communicate with or reach the mainland until their scheduled rendezvous. Days or weeks were spent walking or traveling by sampan along island coastlines, collecting abalone meat and shells, and transporting their commercial products back to central camps for processing and shipment to mainland California markets, and eventually on to China. Surprisingly, there is little to no archaeological evidence that these fishermen sustained themselves on local marine resources. After decades of recovery from intensive Native American fishing, Channel Island ecosystems likely teemed with shellfish, fish, and seabirds. California mussels (*Mytilus californianus*), giant owl limpets (*Lottia gigantean*), and gooseneck barnacles (*Pollicipes polymerus*) would have been exceedingly abundant and found in the same habitats as black abalone. Nearshore and kelp-forest fish such as rockfish, California sheephead, and surfperch, all species that Chinese fishermen pursued elsewhere in California, were abundant. Despite excavations at logistical processing camps and large basecamps, there is no evidence that Chinese abalone fishermen used locally available resources to provision their foodstuffs. Instead, products were transported to the islands, often in brown-glazed stoneware jars and glass containers that probably contained essential ingredients employed in traditional Chinese cooking such as ginger, pickled vegetables, preserved duck eggs, salted fish and meat, tofu, vinegar, oil, and oyster and soy sauce, which supplemented bags of rice.

Even while camping at remote island locations, the Chinese fishermen consumed imported foods using the same traditional rice bowls popular in their homelands and in Chinatowns throughout the American West.

These bowls were more than tablewares and utilitarian dishes: they acted as powerful symbols of identity and personhood. Decorations on several of the rice bowl fragments from Point Bennett, for example, include depictions of bamboo, a plum tree, and a rock. All three can survive harsh winters and other environmental stressors and symbolize strength and longevity. Conscious or unconscious, these symbols and affirmations of strength were especially important for Chinese fishermen as they toiled for months at a time on the isolated Northern Channel Islands, attempting to carve out a livelihood in a country that pushed them to the social, economic, and geographic margins.

Similar patterns have been identified at mainland U.S. Chinatowns, where nineteenth- and twentieth-century immigrants retained traditional diets and consumption patterns (Collins 1987; Diehl et al. 1998). Such patterning has led many archaeologists, historians, and other researchers to conclude that Chinese immigrants actively and stubbornly resisted acculturation as a means of maintaining traditional Chinese lifeways (for discussion, see Voss 2005:426–429). The artifact assemblage at Johnsons Lee and Point Bennett paints a more complex picture, however. As well as using traditional vessels and tablewares, Chinese fishermen consumed Euro-American foods and products. The recovery of an Ellwood Cooper olive-oil bottle from Point Bennett and sauce and spice bottles from Johnsons Lee suggests a complex mix of acculturation and the maintenance of traditional practices.

A major contributor to the success of the Chinese in the commercial fishing industry was their ability to negotiate across ethnic boundaries and build partnerships both within and outside their communities. Newspaper accounts and Fish and Fisheries Reports (e.g., Collins 1888) document that Chinese fishermen worked with and for local Euro-American merchants. As reported in Santa Barbara's *Daily Independent* on October 27, 1883, for example, Chinese fishermen were employed by Rogers Brothers of Santa Barbara and were sent to San Miguel Island "for abalone, seal skins and oil." Chinese fishermen often relied on Euro-American schooners for transportation to and from the islands and for the reprovisioning of island basecamps. They secured fishing rights by entering contractual agreements with island ranching families on Santa Cruz and Santa Rosa Islands. All of this suggests that Chinese fishermen and merchants were deeply and intricately connected to Euro-American economic and social systems. No small part of their success was due to their ability to navigate new social and economic systems while maintaining cultural and fiscal connections with their homelands and with other Chinese immigrants in California.

Their integration stopped short of transforming some traditions, however. The recovery of the remains of a Chinese shoe, hand-stitched with a leather sole and likely cloth body, suggests that, even far from the mainland, Chinese fishermen retained their traditional dress. A young Canadian traveler, William Perkins (1964:319), described these garments when he observed "two hundred fresh arrivals from the Celestial Empire" in the early years of the California Gold Rush: "They were mostly dressed in the national costume, peticoat trowsers reaching to the knees, big jackets lined with sheep or dog-skin, and quilted, and huge basket hats, made of split bamboo. The lower part of their legs are encased in blue cotton stockings, made of cloth, and with soles fully an inch in depth."

Despite external pressure, Chinese fishermen held tightly even to some harmful practices. The recovery of three opium lamp fragments from Johnsons Lee demonstrates the importance of opium smoking, even on the remote Channel Islands. Chinese fishermen must have carefully packed, transported, and cared for fragile opium lamps; likely they had little choice and were bound by physical and psychological addiction. Regardless, for many Chinese immigrants, opium was a leisure activity, not a vice, used for relaxation, relieving emotional stress, enhancing work performance, and medicinal purposes (Wylie and Fike 1993:300). Opium smoking was fraught with negative connotations and considered a depraved, immoral activity in Euro-American Victorian society, but Chinese fishermen clung to this habit to pass the time, relax, and endure the arduous work at island basecamps.

Using the broadest of strokes, the archaeological record paints a picture of hardship, isolation, and strength for Chinese abalone fishermen on the Northern Channel Islands. The only known firsthand account of their island activities was published in 1897 and written by Ninette Eames, who spent three weeks visiting San Nicolas Island with a party of six. After landing on the island, the group encountered piles of discarded abalone shells and a Chincsc fishing cabin:

> "Bless me! [Gastropods] by the ton!" he [one of the men] cried, delightedly, making a dive for one of the numerous shell mounds now everywhere in sight, and beginning to dig like a beaver.
>
> To our surprise we now perceived a small, weather-stained cabin a few rods beyond. Running to it, we flung aside the burlap door curtain, and sniffed disgustedly.
>
> "Chinamen's camp! Opium, dried fish, and their usual accumulation of filth!"
>
> Leroy pointed to ill-smelling blue garments and wooden-soled shoes stacked beside a tea box on which were heaped a motley array of chopsticks, pipes and green rice bowls swarmed over with flies and ants.
>
> Outside the cabin were piles of beautiful abalone shells of larger size and richer tints than any I had ever before seen.
>
> "These fellows must be off shell gathering. Father says they take away millions of abalones from San Nicolas every year. Have you ever seen the meat dried? It looks like the hoof of a colt. I hope they have a boat, so we can borrow it." (Eames 1892:662)

Later, while walking along the coast, Eames and her companions encountered a group of Chinese fishermen gathering black abalone:

> We heard a succession of scraping blows, and over the sandstone cliff discovered a dozen Chinamen making use of crowbars to pry off the abalones from the rocks. Their large baskets were nearly filled with the mollusks already secured. These shell gatherers fought every foot of their way over sharp, slippery stones, with the heavy swells of the sea disputing their advance. They loosened the abalones by means of the powerful leverage of the crowbars, deftly catching them as they fell before the hungry surf sucked them under. The work was extremely laborious and attended with more or less danger.(Eames 1892:665)

In many ways, Chinese fishermen were caught between and within multiple worlds. Chinese immigrants in California were commonly viewed as oddities, often with suspicion and disgust. But their success depended (at least partially) on their ability to capitalize on business arrangements in both the Chinese and Euro-American communities. Most came to California with plans of returning home to their families and friends, dreams that hinged on their ability to successfully negotiate the economic and social hardships of their American lives. The toil of Chinese abalone fishermen working on isolated island coastlines for months at a time in the hope of building a better life 6,000 miles and an ocean away echoes the dreams of countless immigrants worldwide, past and present.

Slowly, however, their grip on the commercial abalone industry (a fishery that they created) began to wane in the 1880s. By this time, the transcontinental railroad was complete and economic depression had spread from the East Coast to the West. California's gold fields had largely dried up, and the temporary boom of the Civil War economy had passed. Anti-Chinese movements spread across the country, and hostility toward Chinese immigrants by white America reached a fever pitch. In the next chapter, we see how Chinese abalone fishermen maintained a tenuous hold on the industry until about 1915, despite increased competition and the implementation of racist legislation. We explore claims that these laws and taxes were designed to protect California's marine food supply from Chinese fishermen who indiscriminately harvested abalone with little regard to the health and long-term viability of the fishery. After the last of the Chinese abalone fishermen lost control and walked away from the fishery, the industry spiraled into rapid decline and lasted for less than 80 years.

CHAPTER 7

A March toward Exclusion

Twilight of the Chinese Abalone Fishery

The driving force behind immigration to the United States for most Chinese migrants in the mid-nineteenth century was to escape the economic depression and ethnic discrimination that gripped their homeland. Initially, America offered Chinese émigrés shelter, to some degree, from both of these. Growing economic prosperity in California and the American West allowed many to carve out successful livelihoods. The mountains offered gold and mining jobs; the coast, fishing; the railroads, steady employment; and an exploding population, opportunities in business and development. Racial and ethnic tensions certainly permeated American society, especially as new groups from Europe, South America, Asia, and around the world transformed the social and cultural landscape of the young nation, but Chinese immigrants found and created economic opportunities. They opened businesses, bought property, and initiated partnerships with white Americans. When times are good, racial tensions are often kept in check, left to simmer below the surface. When hardship sets in, however, and people struggle to support themselves and their families, racial tensions can boil over as people point fingers and look for someone to blame. It is often easiest to impugn outside groups, those who do not conform to social norms or display some outward expression of "otherness"—such as dress, language, skin color, and food preference. Chinese immigrants were easy targets.

Economic Depression and the Seeds of Hate

By the 1870s, a variety of factors coalesced to push America into a nationwide economic depression. Ironically, the downturn was prompted by the solving of two of the country's most pressing problems. In 1869, California governor Leland Stanford drove a ceremonial golden spike at Promontory, Utah, signaling the official completion of the transcontinental railroad. Many believed that the culmination of this engineering and logistical feat would bring economic prosperity to California and the American West. The railroad would allow agricultural and manufactured goods from the West Coast to be brought to market along the more populous eastern coast. The unintended consequence, however, was that cheap manufactured goods flooded the West, and new factories in California could not compete with the established sweatshops, mills, and factories of the East Coast (Chang 2003:117).

At about the same time, Robert E. Lee surrendered the last major Confederate army to Ulysses S. Grant, bringing an end to five years of bloody civil war in 1865. Thousands of former Union and Confederate soldiers headed back home to search for work. Many who could not find employment in the East and South headed to California. Before the completion of the transcontinental railroad, only the most adventurous or desperate made this journey, but now it was open to nearly anyone. Job seekers flooded the West by the thousands; they were not only former soldiers but also European immigrants via New York (Ellis Island after 1892) and other major East Coast cities. Although San Francisco and Los Angeles were rapidly expanding, they were still unable to meet these labor demands. To make matters worse, the completion of the railway and the collapsing gold mining industry left thousands out of work, not only laborers but also the merchants and entrepreneurs who built businesses supplying railway construction, mining, and prospecting (Chang 2003:117).

Chinese workers found ways to succeed, even in the face of a saturated labor market. By 1870, 10 percent of farm workers in California were Chinese; by 1884, this number increased to 50 percent; two years later, it stood at nearly 90 percent (Chan 1986; Chang 2003:73; Voss and Allen 2008:11). Others pursued mining to the east and north, in U.S. territories that later became Nevada, Idaho, Montana, and Colorado (Chung 2011; Valentine 2002). Many Chinese secured employment in factories around the country such as garment plants, canneries, and food processing and manufacturing centers. At the same time, Chinese businesses were rapidly spreading and expanding to include not only laundries and restaurants but also retail stores, home delivery services, factories, domestic work, entertainment houses, and medical and law offices (Chan 1991:33–35; Chen 1981:55–61; Chang 2003:48–49; Voss and Allen 2008:11).

Hard times were felt not only by laborers but also by companies, factories, and businessmen. In an effort to cut costs and maintain or increase their profits, many businesses turned to Chinese labor. In some cases, the Chinese were willing to work longer hours for less money than their Euro-American counterparts. Chinese workers at times crossed picket lines and acted as strikebreakers (Chang 2003:118). Reverend John Todd, a clergyman from Massachusetts, traveled through California in the mid-1860s and described an incident in which Chinese workers served as scabs during a factory strike by an Irish union in San Francisco:

> "In the factories of San Francisco they had none but Irish, paying them three dollars a day in gold. They struck, and demanded four dollars. Immediately their places, numbering three hundred, were supplied by Chinamen at one dollar a day—and superior workmen they are. So it will be all over the country; for all over the country they will come, and be welcome" (Todd 1870:282).

Todd's prediction that Chinese workers would be welcome throughout the country may have been true for some wealthy business owners looking to capitalize on cheap labor, but for the majority of white Americans, the Chinese threatened their way of life. Moreover, Chinese immigrants themselves fought for fair labor practices and living wages even though they were not allowed to join white unions. Chinese laborers organized strikes, sought legal recourse against exploitative working conditions, and cooperated with white labor unions (Baxter 2008:33; Chan 1991:81–83).

Regardless, the Chinese were blamed for the nation's economic woes and were seen as a hostile threat to the "American workingman." Working-class white America complained that Chinese laborers happily worked longer hours for less pay, purchased essential goods in exclusively Chinese businesses, and segregated themselves from "American" society. Following similar arguments to those that resulted in the foreign miners' tax of 1853, many claimed that the Chinese were draining the country of its wealth by sending remittances back to their families in China and continuing to purchase imported Chinese food and products. In short order, white America disregarded the contributions Chinese labor and businesses had made to the young country. It was less than 20 years since parades and gatherings honored Chinese immigrants in San Francisco and the *San Francisco Daily Alta California* (May 1851) predicted that "China Boys will yet vote at the same polls, study at the same schools and bow at the same Altar as our own countrymen."

This is not to say that racial tensions had not lingered since the Gold Rush and permeated social classes very early in California. Governor Stanford vowed during his inaugural address of 1862, for example, that "the settlement among us of an inferior race [the Chinese] is to be discouraged" (Wu 1972:106). It was not until the 1870s, however, that violent anti-Chinese movements erupted and spread like wildfire. This was especially true for cities like San Francisco where, by 1880, there was one Chinese and two white Americans for every available job (Chang 2003:117). Chinese laborers were loathed and feared by white America; they were easy scapegoats, and anti-Chinese rhetoric served to mask underlying tensions between business owners and laborers and between American-born and foreign-born nationals, regardless of race (Saxton 1971; Voss and Allen 2008:11).

Anti-Chinese clubs formed throughout California, holding public meetings, recruiting support, and pressuring politicians to drive Chinese residents out of their cities. The Cubic Air Ordinance of 1870, for example, which was passed by the city of San Francisco (and quickly adopted by the state), required that residences provide at minimum 500 cubic feet of open space for each occupant (Gibson 1877:361–362; Wu 1972:13–14, 65). Although this law was not overtly racist, its enforcement certainly was. Violators in Chinatown were targeted while infringements by white residents were overlooked. Reverend Otis Gibson (1877:361–362), a California missionary in the mid-1870s, described San Francisco police officers "arresting [Chinese] in their beds, in the middle of the night, and driving them like brutes, in droves of forty and sixty, to prison, because found in rooms with less than five hundred cubic feet of space to a man." Ironically, Chinese violators were locked in cells "twice as crowded as were the rooms from which they were taken" (Gibson 1877:362). The fine was set at $10, and if the Chinese were unable or unwilling to pay, their queues were cut off—the infamous "Queue Ordinance" (Wu 1972:14). This was a blatant strategy to humiliate and demoralize Chinese men. Shaving one's queue was an act of treason in their homeland and made a return to their families and birthplaces virtually impossible.

The Chinese were especially susceptible to targeted, racist laws because they were essentially powerless in the American judicial system. Even though the Fifteenth Amendment to the U.S. Constitution was ratified and adopted in 1870, prohibiting state and federal governments from denying the right to vote to any citizen based on "race, color, or previous condition of servitude," Congress deliberately withheld Chinese immigrants' right to naturalize, making it impossible for them to become citizens and engage in the electoral process (Daniels 1988:43). Male children of Chinese immigrants, born in the United States, would one day be eligible to vote, but their immigrant parents were declared "aliens ineligible for citizenship." As pressure mounted for politicians to address the "Chinese problem," discriminatory laws and taxes became commonplace and saw little opposition. A $2 tax per quarter, for example, was levied on laundries with a horse-drawn vehicle and $15 per quarter for foot deliveries—practically all Chinese laundries fell into the latter category. The sidewalk ordinance made it a criminal offense to carry a shoulder pole with baskets on each end in San Francisco (Mark and Chih 1982:33; Wu 1972:14). These measures targeted successful Chinese laundry businesses, which were welcomed and celebrated during the height of the Gold Rush but reviled after its decline.

At the same time, violence against Chinese immigrants became commonplace and increasingly heated. Anti-Chinese rhetoric created a dangerous climate for all Chinese immigrants, men and women, young and old. Reverend Gibson (1877:360) recounted one such ugly and unprovoked incident in San Francisco:

> On Sunday afternoon while a small Chinese youth was pursuing the even tenor of his way along Clementina Street near Eighth, he was suddenly set upon by a crowd of hoodlums,...the entire gang of valiant young Americans pitched into him with rocks, clubs, and other articles of warfare, cutting a large gash in the back of his head, and finally leaving him senseless upon the pavement. He was picked up by some of his countrymen and carried into a wash-house where his wounds were dressed. During this outrage a crowd of interested, full-grown male and female hoodlums stood rubbing their hands over the fun which they did not try to interrupt.

The Chinese had few options to curb the growing violence or to seek justice against their assailants. As we saw in Chapter 4, Chinese immi-

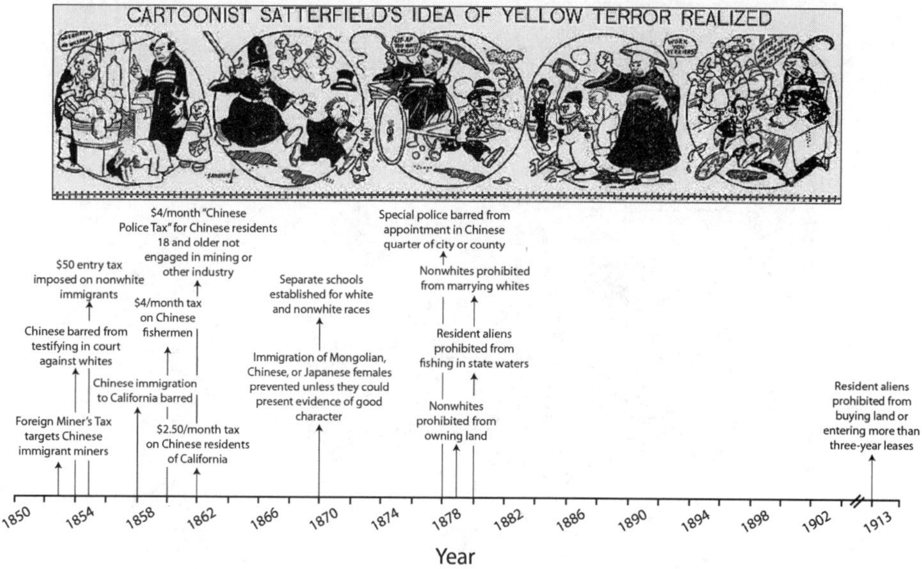

FIGURE 7.1. Timeline of major anti-Chinese legislation and court decisions in California. *Sources:* Baxter (2008), Chan (1991), Chang (2003), Hsu (2000), Wu (1972). *Inset:* Satirical cartoon published in the *Milwaukee Journal* (May 26, 1904) by Bob Satterfield, depicting the artist's vision of life for white Americans if they were treated in the same way as Chinese immigrants.

grants, as people of color, were barred from testifying in court against a white person (Chen 1981:45). While discrimination against the Chinese was widespread in the United States at this time, along with discrimination against other immigrant groups and people of color, laws enacted in the Pacific Coast states tended to be the most restrictive and punitive against the Chinese (Voss and Allen 2008:10; Figure 7.1). In many East Coast and southern states, for example, people of Asian descent were allowed to intermarry with non-Chinese, which was prohibited in western states (Chang 2003:110–113).

In this climate of economic depression, hate, discrimination, and intolerance, a 31-year-old bankrupted Irish immigrant, Denis Kearney, gathered an ardent following in San Francisco (Chang 2003:125–126). After losing everything as a prospector in the mining industry, Kearney became a popular orator, advocating for the overthrow of corporate monopolies, government corruption, and the Chinese. Disgruntled workers, desperate for someone to blame, quickly took to his message, and anti-Chinese rhetoric became a popular rallying cry for Kearney and his supporters. The

"THE FIRST BLOW AT THE CHINESE QUESTION"
The Wasp [San Francisco] 2 (1878): 289.

FIGURE 7.2. One of many cartoons and propaganda circulated by the Workingmen's Party targeting Chinese laborers and fanning the anti-Chinese movement in California. (Published in *The Wasp*, San Francisco, 1878)

Chinese were ideal targets because they could be readily identified and were susceptible to retribution with very little legal recourse to defend themselves. Kearney took to ending each of his public speeches with the slogan "The Chinese must go!" (Chang 2003:126). Kearney argued that "the Chinese laborer is a curse to our land, is degrading to our morals, is a menace to our liberties, and should be restricted and forever abolished, and 'the Chinese must go'" (Sandmeyer 1991:65), a message he personally spread from coast to coast. Many of Kearney's supporters were of Irish descent, and in 1877, they formed the Workingmen's Party of California and elected Kearney their president (Figure 7.2). The irony of this was lost on Kearney and his followers: just three decades before, Irish émigrés faced widespread discrimination, and Help Wanted signs often included the acronym NINA, No Irish Need Apply (Pan 1994:93). The influx of Irish laborers, less than a generation earlier, constituted the same threat as the Chinese for the Workingmen's Party.

The Workingmen's Party became the face of the anti-Chinese movement and was successful in terrorizing Chinese immigrants and pressuring government officials to address the "Chinese question" by passing discriminatory laws. The party organized rallies against the Chinese, boycotted factories and businesses that hired Chinese labor, and orchestrated violence against innocent Chinese throughout California. In one particularly ugly and violent incident, a group of armed white men, spurred by orders from the Workingmen's Party, stormed a cabin in Chico, California, on March 13, 1877, and shot to death five Chinese farmhands whom the landowner refused to fire (Gyory 1998:94). The assailants poured oil over their murdered victims and burned the bodies. Just two years after the party formed, it lobbied the California legislature to make it a criminal offense to hire Chinese employees on public works (Wu 1972:14, 69). Although the Chinese could work for themselves, it became virtually impossible for them to find other work anywhere in the state. Chinese hired hands were forced to leave California, going back to China if they could afford it or to other states or territories if they could not.

Unfortunately, the anti-Chinese sentiments and legislation in California and around the country in the 1870s were just the tip of the proverbial iceberg. Life for Chinese immigrants in America would only get worse, and state-sponsored racism would expand to federal legislation, passed by the U.S. Congress. It was this legislation that spelled the end for Chinese abalone fishing in southern California and, ultimately, brought about the wholesale collapse of commercial abalone fishing along the U.S. Pacific Coast.

Anti-Chinese Discrimination and the California Fishing Industry

The same motivations that drove the anti-Chinese movement in California and throughout the nation spurred legislation to curb Chinese dominance in Pacific Coast commercial fisheries. Like the laundry business, commercial fishing in California was largely an industry "discovered" and expanded by Chinese immigrants, along with Italians, Portuguese, British, and other émigré fishers. While a variety of ethnic groups tried and (many) failed at gold mining and other prospecting, and the economy in California spiraled into stagnation, the market for fish, shells, and other marine resources remained robust. Many Californians viewed these resources as the property of California or the nation, and Chinese fishermen—classified as resident aliens ineligible for citizenship—were seen as profiteering on the collective resources of "true Americans" (see McEvoy 1986:76).

The first attempt to restrict the Chinese fishing industry came in 1860 with the passing of a license fee of $4 per month for every Chinese fisherman (Wu 1972:12). This tax targeted the large net fisheries in San Francisco and Monterey Bays but was equally applicable to all fishing operations in California, including the abalone industry on the Northern Channel Islands. After four years, the fee was revoked and deemed unethical. Though we can only speculate, it is unlikely that such a tax would have been rescinded at the height of anti-Chinese discrimination in later decades.

During the following decades, the United States and California governments became increasingly concerned over the status of Pacific Coast fisheries and the potential impacts of growing commercial (and largely unregulated) exploitation. Catastrophic declines had been felt in many important commercial fisheries along the East Coast due to overfishing and industrial pollution, and officials wanted to avoid a similar scenario along the West Coast. Commercial hunting for pelts and oil was depleting sea mammal populations. Sea otters had already experienced catastrophic declines along the Pacific Coast, and shore whaling had nearly exhausted its prey by the mid-1870s (McEvoy 1986:75). Salmon canneries along the Sacramento River were expanding quickly and having significant impact on the long-term viability of the fishery (McEvoy 1986:71–72), and many other riverine fisheries were being heavily exploited.

In response, California created a State Board of Fish Commissioners in 1870 to conserve and protect the state's fisheries. The next year, the U.S. Congress established the U.S. Commission of Fish and Fisheries as part of the Smithsonian Institution (Bentz and Schwemmer 2002:148; McEvoy 1986:101–102). In 1879 and 1880, David Starr Jordan, a renowned ichthyologist and the future president of Stanford University, conducted a survey of the Pacific Coast fisheries as part of the tenth U.S. census to assess the causes of catch declines of important commercial species in California. For the most part, Jordan found Pacific Coast fisheries to be underexploited and full of economic potential: "Except for the salmon fisheries of the Sacramento and the Columbia, and the ocean fisheries in the immediate neighborhood of San Francisco, the fisheries of the Pacific Coast exist only as possibilities. For the most part only shore-fishing on the smallest scale is done, and no attempt is made to discover off-shore banks, or to develop them when discovered" (as cited in McEvoy 1986:66). There was still considerable concern, however, over the state of nearshore and inland fisheries. Jordan, falling in line with the anti-Chinese sentiment of the time, targeted Chinese fishermen as the primary source of fishery declines in California in the 1870s and 1880s. Shortly after Jordan's report on the Pacific Coast fisheries for the census, his handpicked protégé, N.B. Scofield, continued the attack of Chinese fisheries by releasing a series of reports filled with anti-Chinese rhetoric (McEvoy 1986:101). Scofield claimed, for example, that Chinese shrimp fishermen were repetitively violating state laws by targeting smelt and other small fish that acted as important prey species for other fisheries. Patrol agents of the California State Board of Fish Commissioners and its chief deputy found no evidence of such violations, however (McEvoy 1986:103). Local newspapers and Euro-American fishermen took to blaming any perceived declines in fish stocks on Chinese overexploitation, and eliminating Chinese fisheries became a state priority in the last quarter of the nineteenth century.

By 1880 and Jordan's fisheries survey, Chinese fishermen had been harvesting black abalone from intertidal habitats along California's offshore islands and mainland for nearly three decades. Jordan wrote that their activities had "stripped the whole coast [between Cedros Island and San Diego] of this shell" by 1879 (as cited in McEvoy 1986:82). He noted significant impacts to abalone populations along the Channel Islands by 1888 (at least 32 years after Chinese abalone fishing began) and claimed that abalone along the Monterey coast were "nearly exterminated" by 1889.

Chinese junks became visible representations of East Asian dominance in the fishing industry, and newspapers derided their concentration in bays, estuaries, deltas, and nearshore habitats throughout the state. Jordan quipped that because of the junks along the Sacramento River, "the whole air and look was decidedly foreign, and I might say oriental" (Collins 1888). The Mexican government became concerned about Chinese abalone fishing in Baja as fishermen expanded their operations south, and in 1879, it assessed a $60 annual license fee on each Chinese junk

(Bentz and Schwemmer 2002:149). Junks in San Diego were considered alien vessels and were subject to exorbitant taxes on cargo, entry fees, and certificates. In the midst of these troubles for Chinese fishermen, bigger problems for Chinese immigrants were brewing in Washington, D.C.

The Chinese Exclusion Act and the Demise of Chinese Abalone Fishing

In February 1881, California senator John F. Miller introduced a bill in Congress that would bar Chinese immigration to the United States for 20 years (Chang 2003:130). In his supporting arguments, Miller raged that Chinese labor was destroying white America and that racial discrimination was both necessary and sensible. Even though many of Miller's colleagues supported the bill, President Chester Arthur vetoed it, arguing that Chinese labor had been a pivotal part of the American economy, especially in the railroad, agricultural, and manufacturing sectors (Gyory 1998: 224–244). Arthur's stance was probably heavily influenced by diplomatic relations with China and fears about American ports in China being shut down by the Qing government (Chang 2003:130). In 1868, China and the United States signed the Burlingame Treaty, in which China granted significant economic incentives to American merchants hoping to monopolize the potentially lucrative Chinese market (Chen 1981). The treaty was amended in 1880 and gave the United States the ability to regulate, limit, or suspend Chinese immigration but not to prohibit it.

Protests against Arthur's veto spread throughout the American West and California, championed by the Workingmen's Party. Angry mobs hung Arthur's likeness in effigy and marched in the streets (Chang 2003:132). Another California politician, House of Representatives member Horace Page, stepped in and introduced a bill that excluded Chinese immigration for a period of 10 years but targeted only Chinese laborers, skilled or unskilled. Chinese merchants, teachers, students, and their servants would be allowed entry. No doubt fearing public backlash, President Arthur signed the bill into law on May 6, 1882, passing the infamous Chinese Exclusion Act. Not only did the act restrict immigration, but Article 14, passed in 1884, stated that "no state court or court of the United States shall admit Chinese to citizenship." Thus began eight decades of widespread institutionalized racism in the United States, with the first and only immigration act to specifically designate an ethnic or racial group for exclusion from the United States (Figure 7.3).

THE ONLY ONE BARRED OUT.
ENLIGHTENED AMERICAN STATESMAN.—"We must draw the line *somewhere*, you know."

FIGURE 7.3. Satirical cartoon published in *Frank Leslie's Illustrated Newspaper* (April 1, 1882, vol. 54, no. 96) depicting a Chinese man being barred entry into the "Golden Gate of Liberty."

Rather than calming anti-Chinese sentiment and pacifying Euro-American laborers, the passing of the Chinese Exclusion Act inflamed fanatical anti-Chinese proponents. Once the borders were closed, the goal became to rid the country of Chinese interlopers—a period that has become known as "the Driving Out" (Chang 2003:132). Many Chinese communities throughout the American West endured a state of terror marked by anti-Chinese rallies, murders, and violent evictions of Chinese from their jobs, homes, and communities (e.g., Chan 1991:50–51; Chang 2003:132–135; McCunn 1996:48–51; Storti 1991).

Even legislators took up the cause of driving the Chinese out of America. The Scott Act was passed in 1888, preventing Chinese who had temporarily left the country, many to visit their families back in China, from returning (Wu 1972:16, 82–86). This meant that more than 20,000 Chinese, with legitimate rights to reentry, and 600 Chinese who were already in transit back to their businesses, jobs, and lives in the United States, were denied permission to return. After 10 years, the Chinese Exclusion Act was extended with the passing of the Geary Act in 1892. Chinese immigration was suspended for an additional 10 years, and Chinese residents had to produce a certificate proving their legal residence in the United States before the closing of the border within one year or risk deportation (Wu 1972:16, 86–90). These laws were especially difficult for Chinese abalone fishermen. Fishermen, considered laborers under the Chinese Exclusion Act, could not immigrate to the United States. Additionally, those who sailed into Mexican waters risked denial of reentry or deportation. Many Chinese fishermen, especially around San Diego, were forced to sell their vessels to Euro-American outfits, and Chinese fishing activities precipitously declined after 1892 (Bentz and Schwemmer 2002:149).

The combined forces of anti-Chinese discrimination, exorbitant taxes targeted at Chinese operations, and exclusionary legislation triggered an abrupt decline in Chinese domination of the abalone industry (Braje, Erlandson, and Rick 2007:119). Japanese and Euro-Americans quickly stepped in to fill the void, and rather than focusing on intertidal black abalone, they launched hardhat diving outfits, expanding the industry to include subtidal species such as red, pink, green, and white abalone. Since Chinese abalone fishermen never exploited these species, and sea otters (keystone predators of abalone) were reduced to near-extinction levels, the fishery for shells and meat was exceptionally lucrative. Japanese divers, initially, were very effective in expanding the commercial sale of both abalone meat and shells, but their success was seen as a threat to the national interests of Italian, Portuguese, and other white Americans (Berryman 1995:144).

Local and state governments responded by passing a series of laws under the pretense of conservation and resource protection. Many coastal California counties passed ordinances restricting abalone harvests to deeper waters in 1900, severely limiting Chinese intertidal black abalone fishing (Cox 1962:77). The California state legislature passed laws that restricted abalone fishing to specific seasons by 1905, and counties around California began making concerted efforts to restrict harvests. By 1910 and 1911, size restrictions were placed on a variety of abalone species, including black abalone at 12 inches (30.5 cm) in circumference, equating to a length of 3.8 inches (9.7 cm); red abalone at 17 inches (43.2 cm) in circumference, 5.4 inches (13.7 cm) length; green abalone at 16 inches (40.6 cm) in circumference, 5.1 inches (13.0 cm) length; and pink abalone at 14 inches (35.6 cm) in circumference, 4.5 inches (11.4 cm) length (*Los Angeles Times*, April 15, 1911). In 1913, export of abalone was banned, and by 1915, abalone drying was forbidden, essentially ending the Chinese intertidal abalone fishery (Bonnot 1930; Rogers-Bennett et al. 2002:99). These restrictions were justified as conservation efforts but were motivated by the desire to end Chinese and Japanese involvement in the abalone industry. The California Fish and Game Commission (1910) freely admitted as much: "The real purpose of the law is aimed at the aliens, Japanese and Chinese principally, who were taking them by the ton without regard to size by the aid of diving suits, removing the meat from the shell in the water, bringing it ashore, where it was dried and shipped either to China or Japan." Enforcement of these laws was rigorous and geared toward catching Chinese and Japanese violators. Additionally, an annual license fee of $400 was levied on abalone fishing, with the goal of reducing the number of Asian fishermen. Newspaper articles reported, however, that enforcement of abalone fishing taxes and size restrictions was exceptionally difficult since much of the fishing was centered on remote locations around the Channel Islands (see, e.g., *Los Angeles Times*, October 13, 1902). Nevertheless, an article in the *Santa Barbara Morning Press* (August 26, 1905) suggests that restrictions and enforcement were effective in ending Chinese involvement in the abalone industry: "Two more members of the Ah family yesterday paid fines at Justice Wheaton's court for having undersized abalones in their possession. They were Ah Hoy and Ah King. Each paid $25, pleading guilty to the charges preferred by patrolman H.J. Ables. They were arrested on San Miguel Island as told in yesterday's press. The Chinamen have announced that they will abandon abalone fishing as the fines imposed with a tiresome regularity since the patrol arrangement of the state fish commission went into effect, are eating into the profits of the business."

Evaluating the stock status of black abalone in southern California during the late nineteenth and early twentieth centuries is exceptionally difficult. There are very few records of abalone sizes, densities, and distributions that date back to this period. The primary data source for evaluating the health of abalone stocks and the fishery during this interval is qualitative and clearly colored by anti-Chinese sentiments and racist agendas. Commercial landings data are not available until the 1950s, after recreational abalone fishing and heavy commercial exploitation had been in place for nearly 100 years.

Racism or Conservation? Evaluating Size Restrictions of the Early Abalone Fishery

There is little doubt that Chinese abalone fishing had significant impacts on the health and structure of the California black abalone fishery. Chinese fishermen were taking and processing thousands of pounds of abalone. The fishery peaked in 1879 at 1,860 tons (Rogers-Bennett et al. 2002:99; Cicin-Sain et al. 1977), and Chinese fishermen were expanding operations from San Diego south to Baja California, Mexico, to replenish deleted stocks by the 1860s (Bentz and Schwemmer 2002:149). There remain, however, important questions about the scale of their impact and whether Chinese fishermen were, indeed, "gather[ing] every available abalone on the coast of southern California" (Croker 1931:60). There is frustratingly little data on the abalone fishery before 1879, however, making it very difficult to evaluate the health of the fishery and the impacts of Chinese fishermen in the first two to three decades of the commercial fishery. Were the conservation measures of the late nineteenth and early twentieth centuries designed to save California abalone from extinction at the hands of "thoroughgoing Chinamen with their usual lack of foresight" (Croker 1931:60)? Or were these minimum size limits, maximum bag limits, taxes, fees, and restrictions an extension of racist legislation targeting Chinese laborers?

Without U.S. Fish and Fisheries data before 1879 on abalone shell and meat yields, we are forced to consult alternative sources to evaluate the health of the abalone fishery during the Chinese years. One of the only sources of information comes from archaeological records. Archaeological sites from this time period are a record not only of human activities but also of the transformation of local marine ecosystems—at a time when southern California fisheries became a critical part of globalized food exports.

Part of my survey and excavation efforts on the Northern Channel Islands included measuring whole abalone shells at as many historical

TABLE 7.1. Measurement Data of Black Abalone Shells from Historical Black Abalone Logistical Foraging Sites and Basecamps on the Northern Channel Islands.

Site CA-	Stratum	Count	Mean Length	Min Length	Max Length	Standard Deviation
Santa Cruz Island						
SCRI-715/H	surface	7	115.0	86	131	14.1
SCRI-853	surface	50	113.8	75	148	17.6
SCRI-889	surface	43	109.4	86	131	10.3
SCRI-906	surface	19	117.6	81	166	24.4
SCRI-909	surface	8	124.1	108	139	10.5
SCRI-917H	surface	27	103.6	72	133	17.3
Santa Rosa Island						
SRI-31/H	surface	32	132.3	109	145	8.6
SRI-109/H	surface/A	94	125.4	75	154	16.0
SRI-91-75	surface	153	114.1	54	149	19.1
SRI-471H	surface/A	268	122.4	53	156	17.7
SRI-795H	surface	16	120.9	88	179	23.2
SRI-796H	surface	9	128.3	110	175	18.2
SRI-798H	surface	27	121.7	87	150	19.7
SRI-803H	surface	35	132.5	95	160	18.5
SRI-805H	surface	54	132.9	80	180	26.3
San Miguel Island						
SMI-237/H	surface	19	150.5	138	169	6.6
SMI-273H	surface	9	151.7	146	169	7.7
SMI-553	surface	58	135.7	92	176	18.4
SMI-558	surface/A	115	127.2	31	162	17.7
SMI-614H	surface/East	106	120.9	47	170	23.5
SMI-654H	surface	21	132.0	107	160	16.9
SMI-655H	surface	64	121.0	79	164	20.2
SMI-676H	surface/A	149	132.4	86	162	14.4

Note: All measurements in mm.

Chinese abalone processing sites as possible. In total, 1,383 black abalone shells from 23 historical Chinese abalone processing sites have been measured. One hundred fifty-four measurements from six sites come from Santa Cruz Island, 688 measurements from nine sites come from Santa Rosa Island, and 541 measurements from eight sites come from San Miguel Island (Table 7.1). The overall average length of black abalone shells from Chinese period sites is 125.8 mm (20.8 mm standard deviation). This is over 28 mm larger than minimum size restrictions instituted by coastal California counties in 1910 and 1911. In fact, not a single site produced mean sizes smaller than twentieth-century black abalone size restrictions.

The smallest averages (103.6 and 109.4 mm) were still between 6 and 12 mm larger than historical size limits, and both came from Santa Cruz Island, where mean lengths are statistically significantly smaller than those from Santa Rosa and San Miguel island sites. This is likely a product of environmental conditions and marine upwelling rather than human impacts, and a reflection of trending abalone size reductions from west to east across the Santa Barbara Channel region.

Of the 1,383 individual black abalone measurements, 134 (9.6 percent) are smaller than minimum size requirements. While approximately 10 percent may seem high, we must take into consideration the age structure represented by the Chinese harvest. The size cutoff between adult and juvenile black abalone has generally been marked at about 50 mm; individuals with maximum shell lengths at or over 50 mm are classified as adults, and individuals smaller than 50 mm are considered juveniles (see Leighton and Boolootian 1963; Miner et al. 2006:109). Only two (0.1 percent) black abalone shells from Chinese abalone fishing sites would be considered juveniles, and only 162 (11.7 percent) shells measured less than 100 mm—double the size of the smallest adult.

The Chinese fishery can also be compared with the ancient Native American fishery for black abalone on the Northern Channel Islands. Archaeological evidence for Native American exploitation of black abalone extends back nearly 10,000 years, and the resource was continuously and intensively fished until the Spanish Mission period began in AD 1769 (Rick et al. 2005). From 1,212 individual black abalone shell measurements at San Miguel Island archaeological sites, mean shell size for seven Early Holocene (11,500–7,500 years before present) deposits was 76.7 mm, for nine Middle Holocene (7,500–3,500 years before present) deposits was 102.5 mm, and for seven Late Holocene (3,500 years before present–AD 1542) deposits was 75.0 mm (Erlandson, Rick, Braje et al. 2008:Table 1). The two largest mean shell sizes come from a 7,300-year-old deposit at SMI-388 (103.0 mm, n = 2) and a 4,600-year-old deposit at SMI-396 (108.9, n = 321). No other site produced an average over 92.1 mm, and 14 sites (60.9 percent) produced relatively small averages between 54.6 and 78.0 mm. These data suggest that black abalone through 10,000 years of Native American harvest were significantly smaller than prey sizes during the historical Chinese fishery (Erlandson, Rick, Braje et al. 2008:Table 3). Even with the presence of sea otters, prime predators of abalone, in local waters, Native groups maintained a harvest of smaller and younger black abalone when compared with the Chinese fishery. The ancient fishery was also maintained in the face of millennia of human population growth, fluctuating kelp forest coverage, and changes in sea-surface temperatures.

Taken together, these data provide vital clues for learning about harvesting patterns and human impacts on the fishery during the Chinese period, but first we must understand the life cycle of black abalone. Black abalone, like other abalone species in southern California, are broadcast spawners, releasing sperm and unfertilized eggs into the ocean at the same time (Cox 1962). Fertilized eggs settle to the sea bottom, hatch, and spend five days to two weeks free-swimming. They then settle on the substrate, usually near other individuals, where they transform into their adult form. They develop a shell and settle into cracks and crevices, where they find shelter from predators. Black abalone grow most quickly during the first five to ten years of their lives, and growth rates are affected by disease, food availability, climatic conditions, and other environmental stressors (Haaker 2001). Sexual maturity is reached at less than two inches (50.8 mm), but black abalone remain hidden until they reach a maximum length of about four inches (101.6 mm). Large black abalone leave the relative safety of crevices and their hiding places under large rocks and congregate in tide pools, where they feed on drift algae transported to intertidal zones by waves and currents (Haaker 2001).

The vast majority of black abalone harvested on the Northern Channel Islands by Chinese fishermen measured over 100 mm (Figure 7.4). Considering that these large black abalone tend to be found on the open rocky shore substrate and that smaller (< 100 mm) abalone are hidden in difficult-to-access cracks and crevices, it seems that Chinese fishermen concentrated their efforts on the large, adult individuals and bypassed the juveniles. This pattern can be best explained in one of two ways. First, since black abalone were hyperabundant in southern California during the mid-nineteenth century, at least initially, Chinese fishermen may have targeted the largest, easiest-to-find individuals for exploitation. Rather than spend time searching for and collecting harder-to-locate juveniles, Chinese fishermen maximized their foraging efficiency by focusing on large adult individuals. In much the same way, market demands for large abalone steaks and, later, for large shells for ornamental display, jewelry, and lacquerware may have driven fishing behavior.

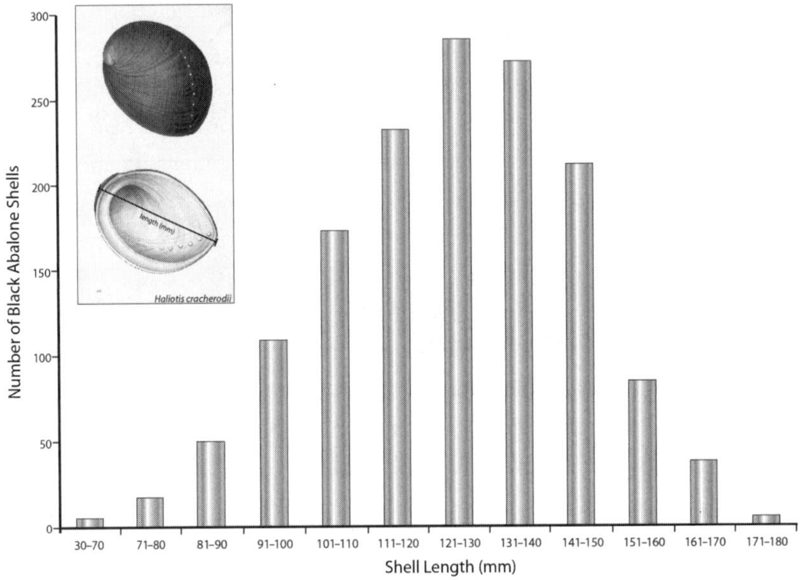

FIGURE 7.4. Measurement distribution of black abalone shells from historical abalone processing sites on California's Northern Channel Islands. *Inset:* Black abalone shell and maximum length measurement. (Open access image via http://commons.wikimedia.org)

I believe the alternative scenario is more likely, however. That is, Chinese fishing for large adult abalone was part of a deliberate and strategic conservation strategy. Chinese fishermen intentionally bypassed juvenile abalone in order to maintain a seed population and preserve the long-term viability of the fishery. Chinese abalone fishing in southern California was initially successful because the abalone fishery in mainland China was heavily overexploited and the industry had collapsed, and California abalone served as a replacement for the Chinese fishery. Since Euro-Americans were originally uninterested in abalone steaks or cuisine, the American abalone market in the nineteenth century was a small, specialty one serving Asian immigrants, primarily in California's Chinatowns. The real profitability of the industry was realized when a large portion of the abalone meat and shells were shipped overseas to meet demand in China. Chinese fishermen in California were well aware of the demographic collapse of abalone in the their homeland and likely took steps to help ensure the American fishery's long-term viability. This would have been critically important since opportunities in gold mining, prospecting, farm work, and other professions were closed to Chinese immigrants as part of the growing anti-Chinese movement. Fishing, and particularly abalone fishing, was one of the cottage industries largely controlled by Chinese immigrants for some three decades. Its long-term preservation was essential for their economic success in California, their ability to send remittances home to their families, and the chance to one day reunite with their loved ones.

The nineteenth-century California abalone industry was also much different from many other commercial fisheries, past and present. Borrowing from economics theory, biologists and marine resource managers have demonstrated that the precipitous decline of economically important commercial fisheries around the globe can be attributed to what has been popularly called "the tragedy of the commons." Fish populations are classic examples of common property resources: they are not owned by any single individual or group. According to the tragedy-of-the-commons model, when individuals operate independently and rationally according to self-interest, they tend to deplete the common resource and negatively impact the interests of the larger group (Hardin 1968). Selfish fishers, then, are locked into a deterministic scenario in which they are both the villains and the victims—without regulation, collapse is inevitable. Fishery collapse due to the tragedy of the commons is not universal (see Feeny et al. 1996), however, especially in many community-based and small-scale fisheries, where implicit or explicit customary laws or regulations prevent individual fishers from maximizing private gains in the interest of the community.

The Chinese abalone fishery was controlled by a relatively small number of merchants and fishermen. Unfortunately, it is impossible to know the exact number, especially in the first 15 to 20 years of the fishery, for which there are virtually no historical documents. U.S. Fish and Fisheries Reports and census and tax records (see Berryman 1995:47–53) suggest, however, that the number of Chinese fishermen did not exceed 25 individuals in major southern California fishing hubs such as San Diego and Santa Barbara. Many of these records are likely inaccurate since some Chinese fishermen lived on their boats, engaged in long-term fishing expeditions, or were infrequently visited by census workers (see Jordan 1887:37), but the records also do not distinguish between abalone and other kinds of fishing. Of the relatively small number of coastal fishermen identified in historical documents, probably only a fraction concentrated on abalone exploitation. A relatively small group of culturally and ethnically connected individuals, then, controlled the abalone fishery and likely made harvesting decisions to ensure the long-term sustainability of the fishery.

This should not be surprising since many nineteenth-century ethnic groups in California developed informal procedures and customary rules for protecting their fishing grounds long before government regulation (see McEvoy 1986:95–100). Chinese shrimp fishermen, for example, recognized one another's proprietorship over designated fishing grounds in the Bay Area without formal regulation (McEvoy 1986:96–97). Italian fishermen organized themselves into "The Fishermen of the Bay of San Francisco" in 1862, primarily to guard against incursion by competing ethnic groups. Many similar coalitions formed along ethnic lines and attempted to mitigate the impacts of market forces by regulating harvests. Fishermen were morally obligated to make socially responsible decisions, often at the expense of profit and to the benefit of competitors (McEvoy 1986:99). This pattern was identified by David Starr Jordan in California during his initial fisheries survey: "Everything is governed by laws which the fishermen have made for themselves" (Jordan 1887:732–733). Rather than operate as individual actors, Chinese fishermen developed fishing strategies that helped maintain the long-term health of the fishery. Its viability during three to five decades of Chinese harvest, focused on a single intertidal, slow-growing species, is perhaps the most powerful evidence to support such a scenario.

In the decades following the twilight of Chinese abalone fishing, the industry, the players, and the health of the fishery fundamentally and forever changed. I begin the next chapter by exploring these changes and tracking the collapse and continued degradation of abalone populations in southern California. For some species, such as black abalone, even our best efforts at conservation and restoration have largely failed, and their continued survival in the wild is of critical concern. But it may be the legacy of Chinese and Native American abalone fishing practices and the archaeological records they left behind that can provide roadmaps for the protection, recovery, and restoration of California abalone.

CHAPTER 8

An Enduring Legacy?

By the turn of the twentieth century, Chinese immigrants' dominance in the California abalone fishery had ended and their participation was hanging by a thread. Racist laws, taxes, and ordinances had forced them not only out of the abalone business but out of much of California's fishing industry. By 1893, for example, only one Chinese junk was still operating in San Diego County, a former epicenter of the Chinese fishing industry (Lundy 1997). A few Chinese abalone fishermen managed to hang on for 10 or 15 more years, partly by entering into contracts with private landowners for abalone fishing rights. On Santa Cruz Island, contracts between the Caire ranching family and Chinese abalone fishermen were in place until at least 1908 (Chiles 2011). Fishermen were allowed to work the island shoreline in exchange for a portion of their catch. With California's ban on exporting abalone in 1913 and the drying of abalone in 1915, however, intertidal fishing for black abalone came to a precipitous end (Bonnot 1930). The decades-long, increasingly severe harassment of Chinese fishermen, who were once vital for helping feed other Asian immigrants essential in building the nation's infrastructure, finally met its end goal.

By 1898, Japanese immigrants filled the void left by Chinese abalone fishermen and managed to thrive by expanding operations to subtidal species. Japanese "sake barrel" fishermen dove for abalone in offshore waters, initially using primitive technologies. Free divers would wear goggles, hold their breath, and search for large red abalone in deep water (Lundy 1997). Between dives, they would rest by clinging to empty imported sake (rice wine) barrels repurposed as floats. Cold waters and the difficulty of

FIGURE 8.1. Japanese abalone fishing camp on San Clemente Island. (Photo published in 1913 in *Popular Science Monthly*)

accessing deep-water species quickly led to the replacement of sake barrel divers by hardhat helmet abalone divers with insulated deep-sea dive suits (Cox 1962:77). With new harvesting techniques came new processing and packaging methods. The Point Lobos Canning Company opened its doors south of Monterey in 1898 and employed three Japanese hardhat divers to churn out production of canned red abalone meat (Lundy 1997).

The fishery, at least in the early years, was incredibly productive. Since adult abalone of subtidal species had enjoyed nearly 100 years of little to no predation pressure (Chinese fishermen could not access deep-water species, and sea otters had been locally extirpated), Japanese fishermen regularly collected 40 to 50 dozen abalone daily (Lundy 1997). Japanese boats, crewed by five men—a diver, a line tender, a kelp cutter, a pump operator, and a captain—would find productive habitat and follow the diver as he walked along the bottom, often through thick kelp forests, searching for abalone (Cox 1962:77–78). Kelp was cut out of the diver's path and kept from entangling his air hose while he gathered abalone from the seafloor in a basket or net bag. Reports from Mendocino indicate that in six hours Japanese divers could collect 2,300 red abalone, and

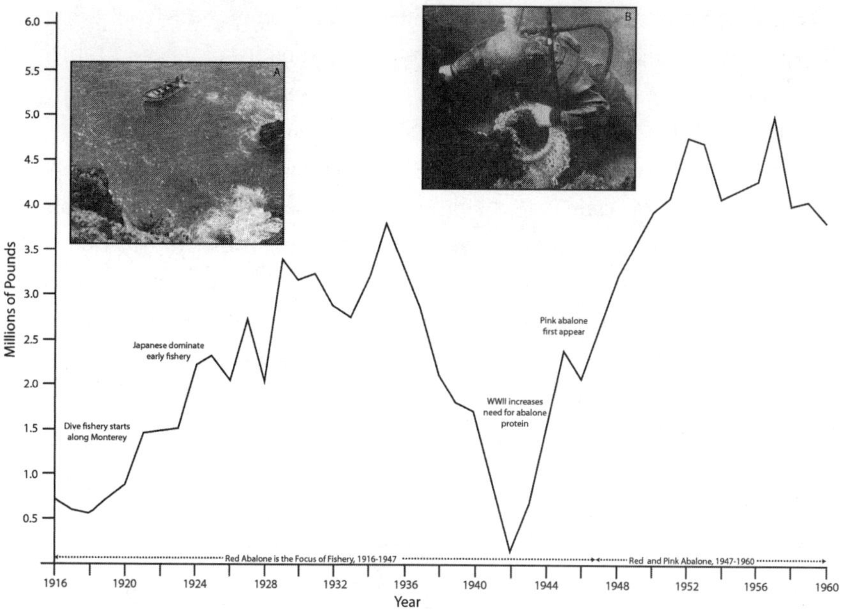

FIGURE 8.2. Approximate total catch of abalone taken during California's dive harvest from 1916 to 1960 (based on Cox 1962:82). *Inset:* (*A*) Japanese abalone divers fishing off California's coast between Carmel and San Simeon, September 1938 (Cox 1962:79); (*B*) abalone diver in hardhat gear gathering red abalone into a woven basket ca. 1957 (Cox 1962:82).

the fishery was equally as productive down into Baja California (Rogers-Bennett et al. 2002:100; Figure 8.1). Japanese divers described the waters along Baja California as teeming with abalone:

> "Abalone was found in abundance too. It was common to find them in layers of twelve or thirteen.... At that time the divers used to catch five to six tons a day" (Estes 1977).

Even with this hyperabundance, red abalone stocks began to decline by 1913, and Japanese divers expanded their takes to include smaller green and pink abalone (Figure 8.2). Prey switching helped maintain the abalone windfall, at least temporarily, as the University of Southern California professor Charles Lincoln Edwards (1913:6–7) observed: "The diver secures a net full of abalones, gives the signal, and the mollusks are hoisted aboard the boat and stowed below. I have seen the diver send the net up, filled with about fifty green and corrugated [pink] abalones, every six or seven minutes. During his shift below the diver gathers from thirty to forty basketfuls, each containing one hundred pounds of meat and shell, or altogether one and one half to two tons." Because the drying of abalone was prohibited, between 1915 and 1920 six abalone canneries opened along coastal California from Monterey to San Diego. Most of these were very short-lived, however; by 1928 only three remained open, and by 1931 only one was still in operation (Cox 1962). Heavy fishing pressure was taking its toll on red, green, and pink abalone populations, and the California Department of Fish and Game took notice. Unfortunately, it is difficult to assess the overall impacts of the fishery across species since, before 1940, only red abalone were identified in commercial landings records.

The first no-take abalone reserve was established between Point Piños and Seaside in central California's Monterey County in 1907. At the same time, periodic closures were enforced at Venice Beach, and conservation advocates and newspapers warned of the dangers of overfishing and the lack of state oversight (see Edwards 1913:11–12). Commercial harvest of abalone was prohibited in southern California from 1913 through 1943, and the fishery concentrated on central and northern California (Haaker 2001:89). Several laws were established by 1930, delimiting abalone fishing districts, creating maximum bag and minimum size limits, outlining legal fishing methods, and establishing seasonal closures (Croker 1931; Lundy 1997; Rogers-Bennett et al. 2002:100). Despite these measures, abalone came under increasing pressure as the recreational fishery exploded in popularity after the 1915 Panama-Pacific International Exposition in San Francisco introduced the general public to the culinary joys of abalone meat (see Chapter 1). Sport fishermen took to the coast by the thousands to hunt for abalone, often with little regard for size restrictions or daily bag limits. A report describing the status of California's fisheries in 1929 asserts, for example: "During the open season, at every real low tide, many hundreds of tourists and ranchers can be seen going over every accessible reef and ledge 'with a fine-tooth comb.' State and county authorities are hard-pressed to enforce the laws on limits and minimum size which are so easily broken by thoughtless people" (Croker 1931:63). At about the same time, the number of dive permits to fish commercially for abalone skyrocketed: 11 were issued in 1928, 27 in 1937, 130 in 1947, 294 in 1954, 505 in 1960, and 880 in 1966 (Cox 1962:80). Permits were required for every member of an abalone fishing crew, including line tenders and boat operators, so hardhat divers are represented by only a small fraction of the permit numbers. Still, the sharp increase in permits suggests intensifying predation pressure by growing numbers of commercial diving outfits. Most of the early permits were issued to Japanese fishermen, whose skills and knowledge of hardhat abalone diving allowed them to dominate the industry for decades (Cox 1962). By the mid-twentieth century, however, large numbers of Euro-Americans were involved. To accommodate the growing fishery, commercial dive operations expanded from an epicenter in Monterey and intensified along the southern mainland coast, south of Point Conception, and along the Channel Islands.

World War II presaged the end of Japanese involvement in California's abalone fishery, as Japanese Americans living in coastal towns and cities were rounded up and sent to inland internment camps (Cox 1962:77). The need for wartime protein, however, spurred the state to open previously closed shorelines to commercial abalone fishing, and landings skyrocketed, for red abalone in northern California and for pink, green, white, and black abalone in southern California (Rogers-Bennett et al. 2002:100). The marine biologist Konstantin Karpov and his colleagues (2000) at the California Department of Fish and Game identified four phases over the 50-plus years between World War II and the complete fishery closure in the mid- to late 1990s. The first phase was characterized by rapid increases in abalone landings, especially red abalone between 1942 and 1945 (Figure 8.3). As reds became less available due to heavy fishing pressure, pink abalone were commercially exploited for the first time, taking the place of reds as the primary abalone species in the California fishery.

The second phase, between 1952 and 1968, appears to be a time of stability for the fishery; landings hovered around 2,000 metric tons per year. This was, in fact, an illusion, as heavy commercial exploitation serially depleted abalone from California waters. Serial decline is a common problem for modern commercial fisheries and occurs when the most desirable and easily accessible species are fished first. When overfishing depletes these, less desirable or more difficult-to-obtain species are added to the fishery and, in turn, undergo heavy fishing pressure. During this second phase, commercial landings looked stable, but when individual abalone fisheries are considered, serial decline by species and fishing locations are

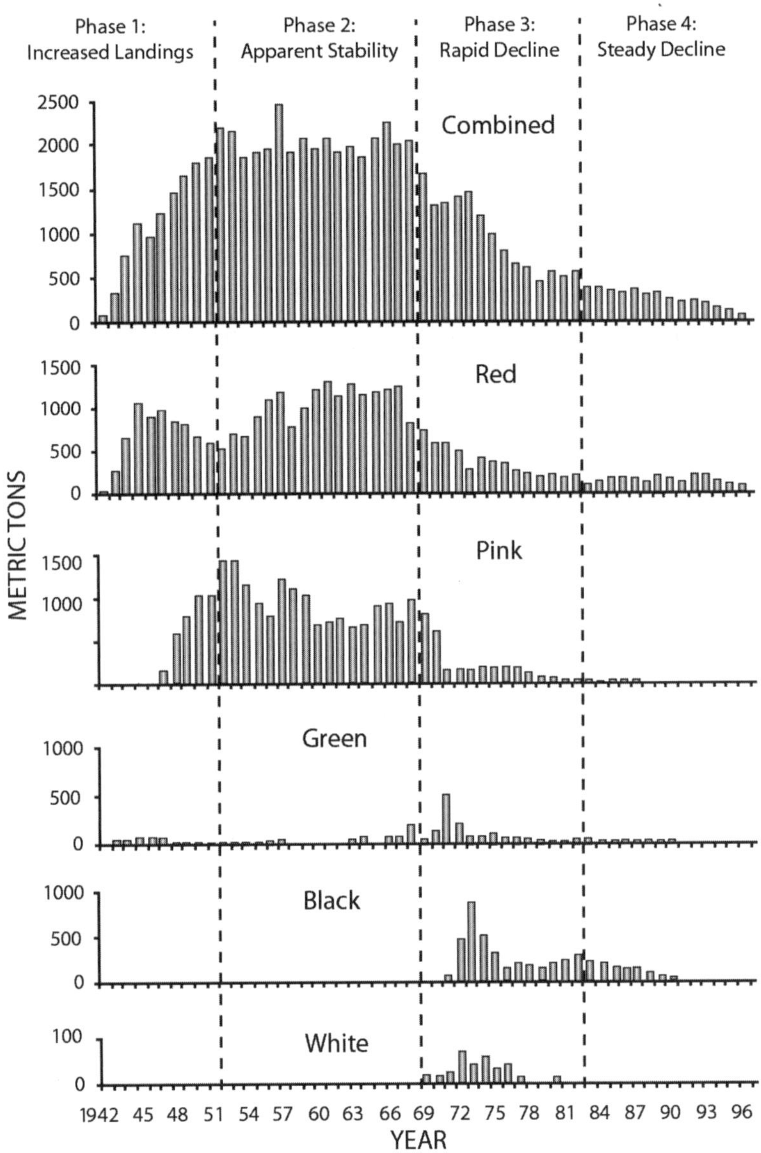

FIGURE 8.3. Summary of commercial landings (metric tons) of the California abalone fishery: combined (*top*), red, green, black, and white abalone (after California Fish and Game Commission 2005:Fig. 3-1; Karpov et al. 2000:Fig. 2).

evident. Overall landings appeared constant because red abalone fishing gradually increased until 1967, as pink abalone landings declined. At the end of phase 2, pink abalone rapidly declined due to increased size limits, but green abalone took their place as size limits for this species were decreased. In addition, new fishing grounds were opened and exploited during this interval and helped to maintain landings at record highs.

Phase 3, between 1969 and 1982, was characterized by obvious signs of overfishing and dramatic declines in abalone landings across all species. The worst declines were of white abalone, which in only 10 years plummeted to less than 1 percent of its historical peak. Every species, even the once robust populations of red abalone, were fished to historical lows. Sadly, green and white abalone continued to be fished even after their California populations dipped below 12,000 and 300 animals, respectively, left in the wild (Rogers-Bennett et al. 2000:101).

In the final phase, despite severe declines in landings across all species save a small but seemingly stable red abalone fishery at historically low levels, closures were not enacted until the mid-1990s. Strict size and bag limits were intended to stabilize the wild population while maintaining the commercial fishery (Tegner et al. 1992), but commercial divers tended to hold both abalone and sea urchin fishing permits. As abalone became less available, fishermen shifted to sea urchins but still took abalone when available (Dugan and Davis 1993). This process reduced the number of adult, spawning abalone and pushed their densities below critical thresholds for recruitment (Levitan et al. 1992; Pennington 1985).

Currently, all five commercial abalone species in central and southern California are depleted. The combined effects of overfishing, expanding sea otter populations, pollution, disease, sea surface temperature changes (e.g., El Niño), and management failures have left California abalone in various degrees of disarray (Haaker 2001:95). The white abalone fishery was closed in 1996, and the species was listed as endangered under the U.S. Endangered Species Act (ESA) in 2001. Today numbers are so low and densities so small that recruitment and reproduction are severely limited. Even with federal protection and careful management, it may be just a matter of time before white abalone are extinct in the wild (Courchamp et al. 2006; Davis et al. 1998). Green and pink abalone were reduced to dangerously low numbers during the commercial and sport fishery, and after nearly two decades of protection and management, they are listed as "species of concern" under the ESA. Their densities and distributions are extremely low, and their recovery is still in doubt (Haaker 2001).

In 1993, the commercial and recreational fishery for black abalone was closed because of declines from overfishing but also because of the devastating effects of a lethal disease, withering syndrome (WS). WS was first identified in tide pools on the Channel Islands in 1985, when marine scientists found shriveled and dying black abalone falling off rocks in large numbers (Haaker et al. 1992; Haaker 2001:94). Population densities of black abalone in protected waters along the Channel Islands often averaged more than 40/m^2 in the mid-1980s (Douros 1985) but plummeted after the appearance of WS. After 16 years of fishing moratorium, careful management, and restoration, in 2009 black abalone were upgraded from "a species of concern" to "endangered" under the ESA. Their abundance was estimated at just 1 percent of historical (1972–1981) levels at sites within the Channel Islands by the late 1990s (Rogers-Bennett et al. 2002), with additional declines as WS spread northward (Raimondi et al. 2002). Recent evidence of some recovery and recruitment has been detected on San Nicolas and Santa Cruz Islands (Butler et al. 2009; D. Richards, personal communication 2012), providing some hope for the future prospects of black abalone in southern California.

Red abalone have fared slightly better than other species in California. The commercial fishery was closed in 1997, and although population levels were reduced to one-tenth of their historical peaks (Haaker 2001:91), red abalone declined the least of all the five abalone species in California. South of San Francisco Bay, recreational and commercial fisheries remain closed, but populations may be rebounding, especially along the Northern Channel Islands (see below). In northern California, a small, highly regulated sport fishery remains open, and data suggest that populations are stable (Karpov et al. 2000; Orensanz et al. 1998). The northern California sport fishery for red abalone is now the only way to enjoy a local, wild abalone steak in the state. From an abalone fishery that survived for at least 12,000 years in California (see Erlandson et al. 2011), this is all that remains. Most Californians who can afford to sample abalone do so at high-end restaurants, paying exorbitant prices for farm-raised abalone steaks grown in tanks along the central California coast.

Historical Ecology: Consulting the Records of the Past to Meet the Challenges of Today

Sadly, the serial collapse of the California abalone fishery is not a unique story. Many commercially important fisheries and marine ecosystems around the world are in a state of crisis, a situation that has been widely recognized for half a century. Despite concerted efforts to reverse the trend of collapsing fisheries using applied ecology, monitoring, and careful management, little has changed (Pitcher 2001:601). Our track record for effective management is abysmal, and year after year we endure fishery collapses and their associated economic and social fallouts. Individuals, families, and entire communities are shaken to the core as their livelihoods, resource bases, and local seascapes erode out from under them.

In 1995, the University of British Columbia marine biology professor Daniel Pauly wrote a concise, one-page article in the prestigious journal *Trends in Ecology and Evolution* titled "Anecdotes and the Shifting Baseline Syndrome of Fisheries." In this seminal piece, Pauly (1995) pointed to the ahistorical approach of modern fisheries science as a major contributor to global fisheries mismanagement and collapse. Pauly argued that every generation of marine scientists tends to accept the stock sizes and species compositions that occurred at the beginning of their careers as the baseline for evaluating subsequent changes. Little regard is paid to earlier periods because many marine biologists and fisheries scientists assume that the data are inadequate. This tendency, Pauly maintained, has led to a slowly shifting perspective on the status of fish stocks and the health of marine ecosystems, which he dubbed "the shifting baselines syndrome." Compounding the problem is the development of more sophisticated modeling techniques, their greater demands on data, the increasing reliance on more recent datasets, and, consequently, an increasingly stringent view of older data as "inadequate." The result is a gradual shift in baselines and the use of inappropriate reference points for evaluating the health and structure of marine systems. Historical amnesia has shifted our perception of what is natural toward more and more degraded ecosystems—a disaster in slow motion that has resulted in the collapse of many of the world's most commercially important fisheries.

Over the last 20 years, Pauly's (1995) call for marine scientists to consider deep historical datasets has led to increased collaboration with historians, archaeologists, economists, sociologists, geographers, and other interdisciplinary scientists. Many of the most prominent scientific journals in the world have published special issues devoted to the topic. New scholars and popular books have helped to educate the scientific community and general public (e.g., Bolster 2012; Braje 2010; Braje and Rick 2011; Jackson et al. 2011; Rick and Erlandson 2008; Roberts 2007; Starkey et al. 2008). Many resource managers and marine biologists now recognize that exclusive reliance on modern ecological and fisheries data can result in a skewed perception of ocean ecosystems and management plans based on already degraded reference points (e.g., Carlton 1998; Dayton et al. 1998; Jackson 1997; Jackson et al. 2001). Increasingly, researchers have turned to marine historical ecology, an interdisciplinary approach that focuses on the long-term evolution of ocean ecosystems, the effects of both natural and anthropogenic changes at multiple time scales, and the dialectical relationship between humans and their environments (Balée and Erickson 2006; Crumley 1994; Gragson 2005). Simply put, marine historical ecology is the integration of marine ecology and history (Rick and Lockwood 2013:46). Archaeology has become a key component of marine historical ecology research, offering a variety of important datasets, including faunal and floral abundances and distributions, baselines and benchmarks that document the structure and function of ecosystems through deep time, prehistoric or historical variability within an ecosystem, and historical and ancient thresholds of human harvest (see Braje and Rick 2013:310). The advantage of a historical ecological approach to the study of physical and biological environments is that it provides a historical dimension that helps answer the question, how did we get where we are today? (e.g., Lepofsky 2009; Redman 1999; Swetnam et al. 1999). Understanding environmental change and ecosystem evolution over multiple chronological scales, from decades to millennia, and spatial scales, from local to global, is essential to learning how and why healthy or damaged ecosystems have evolved to their current states. Only with such long-term data can we develop baselines and protocols for future policy and effective actions in environmental management, conservation, and restoration.

More than just a challenge to be faced by scientists and resource managers, the problem of shifting baselines is one that affects us all. The general public falls victim to the same traps. The things that we perceive as natural today, such as the abundances and sizes of fish and shellfish along our coasts, may be dramatically different from what our parents or grandparents considered natural. Change, whether positive or negative, is measured against our own experiences. If conditions improve, then we view the system as healthy or restored, even if it is just a shadow of its former self. If conditions deteriorate, each new generation considers the new state natural, measuring the health of the system from the yardstick of its own experiences.

Hop a plane to the Caribbean, for example, and pack your swimsuit and snorkel. As you glide through the crystal blue waters above beautiful coral reefs and past colorful groups of schooling fish and parrotfish, it's hard not to feel as though you've entered a Garden of Eden, a pristine seascape. Brilliant colors, clear waters, and a diversity of fish species create the illusion of an island paradise. All you need to do, however, is thumb through pictures from the local big-game sport-fishing competitions and you'll notice a disturbing trend. Photos from the 1950s are of large groupers and other impressive species, a meter or more in length (see McClenachan 2009). By 2007, smiling fishermen are holding small snappers that measure less than a foot long. The big fish are gone, the result of overfishing and ecosystem degradation. To see the coral reefs in all their glory, you need to travel back in time, before the arrival of Columbus, when waters teemed with 300,0000 Caribbean monk seals, 90 million green sea turtles, and a variety of other fishes and sea life, large and small (Jackson 1997). In just the last 50 years, reef fishes have declined by 90 to 95 percent and sea turtles by 99 percent, and the last sighting of a Caribbean monk seal was in 1952 (Sala and Jackson 2011:196). Overfishing, land pollution, coastal development, climate change, and a myriad of other factors are all to blame. Setting goals for conservation and management and repairing the damage will take a herculean effort, but one that remains possible if we use appropriate measuring sticks.

Two key findings of historical ecology are that (1) humans have had significant impacts on the natural world and (2) these impacts extend back into deep antiquity. Any restoration efforts aimed at rebuilding a natural state must take into consideration the deep history of human-environmental interactions and the development of baselines at multiple time scales. What historical ecology cannot do, however, is tell us what our baseline should be. Rather, it provides the most democratic of all ideals:

a choice. Historical ecologists reconstruct ecologies through time, track changes from natural and anthropogenic forcing, and help us understand how we came to where we are today. Historical ecology tells us that even "natural" systems fluctuated through periods of stasis and change. Nature is complex, and deciding how to restore the "natural" is a messy problem, wrapped up in the politics of determining the desired future condition. Conservation and restoration policies are judged not solely on their scientific merits but also on how they influence the public and local stakeholders. The error of our past management strategies, however, was that they were developed while we suffered from amnesia, and they were built from choices we made without the benefit of history and memory. Rather than work from already degraded baselines, from management plans doomed to fail, we can use historical ecology to provide a deeper understanding of how the world once was and a vision of how the world can be.

The challenge of restoring and protecting California abalone populations offers an excellent test of historical ecological approaches. It's hard to argue that our current abalone management plans and policies have not failed. Despite fishery closures, careful monitoring and restoration, and a comprehensive abalone management and recovery plan, many species continue to struggle. Restoration efforts have nearly all been based on historical landings data that extend back less than 50 years, from the mid-1990s back to 1950. Deeper historical data have largely been ignored, even though abundant archaeological evidence is available dating back at least 12,000 years along the California coast. Two Northern Channel Island abalone case studies, in particular, demonstrate the vital importance of historical ecological perspectives. In each of these, the integration of deep historical data can offer a critical roadmap for restoration efforts and future management.

Restoring and Protecting California's Red Abalone Fishery

Despite careful management and monitoring and nearly two decades of fishery closures, there has been little to no sign of improvement for most of California's abalone. Recovery efforts have been complicated by the demographic and geographic expansion of sea otters in central California, ongoing battles to eradicate withering syndrome, and illegal poaching. Further complicating management has been the lack of deep historical baselines (i.e., centuries to millennia) against which to measure the health of California abalone stocks. When they were first commercially exploited after sea otters had been locally extirpated from California and Native Indian populations had been integrated into Euro-American agrarian economies, abalone reached sizes, abundances, and densities that were unprecedented in California. The spectacular yields of the past—in the 1850s with the Chinese and black abalone and in the late nineteenth to early twentieth centuries with the Japanese and red abalone—created unrealistic expectations, all without historical baselines against which to measure a healthy stock under predation pressure (Braje et al. 2009).

Red abalone may be the one remaining bright spot for the recovery of the California abalone fishery and the rebuilding of a commercial fishery. Unlike other species, red abalone have expanded their range and numbers throughout much of southern California, especially along the Northern Channel Islands. The northern California recreational harvest has been stable for decades, and successful fishing regulations (albeit very strict and heavily enforced) are a source of optimism for our ability to construct effective management plans. San Miguel Island, in particular, has seen promising increases in the density of red abalone populations, likely due to strong upwelling and cold-water influx, which made the island the focus of commercial and recreational harvests before the 1997 closure. Compared with the historical declines of the early to mid-twentieth century, red abalone in California appear to be thriving. But are we viewing this "recovery" under shifting baselines and with historical amnesia?

In 2006, a group of former commercial abalone divers requested that the California Department of Fish and Game (CDFG) open a small commercial test fishery along San Miguel Island, citing an amendment of the California Abalone Recovery and Management Plan that allows experimental harvests before stocks are fully recovered. The CDFG formed an Abalone Advisory Group to review the feasibility of the proposal and surveyed red abalone populations along San Miguel. Its plan was to base its determination on modern survey data from San Miguel Island waters. Concerned about the ecological fallout from opening a commercial fishery for red abalone without consulting deep historical baselines, I led a group of archaeologists and marine ecologists in integrating the deep history of red abalone fishing along the Northern Channel Islands.

Our team synthesized all the available evidence of red abalone fishing on the Northern Channel Islands over the last 12,000 years, from the

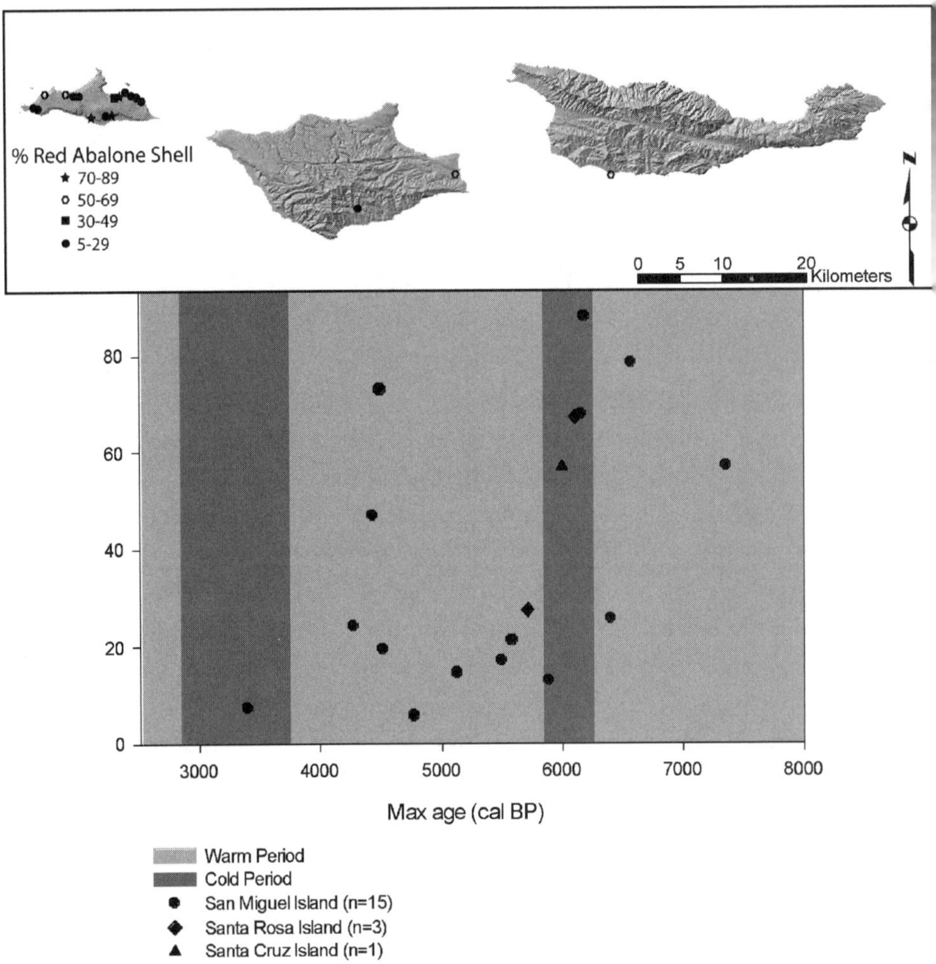

FIGURE 8.4. *Top:* Location map of archaeological sites on the Northern Channel Islands containing abundant red abalone shell (San Miguel Island = 15 sites; Santa Rosa Island = 2 sites; Santa Cruz Island = 1 site). *Bottom:* Red abalone shell percentages from these sites plotted through time. Warm sea-surface-temperature periods are represented as light gray sections, and cold periods are represented as dark gray sections. These data suggest that abundant red abalone were available on San Miguel throughout the last 8,000 years, but became available along islands to the east only during cold-water episodes.

earliest island occupants to modern commercial fisheries. Employing archaeological, paleoecological, historical, and modern catch data, we found that the availability of red abalone across the channel was discontinuous through time. Beginning about 8,000 years ago, archaeological, historical, and ecological data suggest that Native American hunters may have reduced sea otter populations in local watersheds, resulting in exceptionally productive red abalone fisheries (Erlandson et al. 2005). There is no evidence, however, that sea otters were extirpated from the Santa Barbara Channel until the historical fur trade of the seventeenth through nineteenth centuries. One hypothesis is that the reduction of sea otter populations was a deliberate strategy by Native Americans to enhance the availability of economically important abalone, California mussel, sea urchin, and other shellfish. Reduced otter populations resulted in especially high red abalone densities and large sizes along San Miguel Island waters for millennia, independent of local fluctuations in water temperatures. Archaeological deposits on the large islands to the east, Santa Rosa and Santa Cruz, tend to contain abundant and large red abalone shells only during optimal climatic conditions, when sea surface temperatures were colder than normal along the channel (Braje 2007; Braje et al. 2009).

What a historical ecological approach can tell us about the red abalone fishery that modern landings data cannot is that the waters around San Miguel Island sustained intensive red abalone fisheries for at least 8,000 years (Figure 8.4). Through periods of cold and warm waters and high and low productivity, red abalone continued to thrive along San Miguel Island. It is San Miguel Island that provides critical habitat for the red abalone larval production and recruitment that feeds the larger Santa Barbara Channel area. When channel waters are cool and productive, the San Miguel nursery feeds the channel to the east and productive fisheries spring up on Santa Rosa and Santa Cruz Islands. Most importantly, fishermen, resource managers, and ecologists should expect substantial abalone populations to recover around San Miguel Island first, and a better test of the health and recovery of red abalone should be their rebound along island shorelines to the east. Modern surveys of these secondary coastlines may be more effective for evaluating the recovery and health of red abalone communities along the Northern Channel Islands. Historical, archaeological, and ecological data also suggest that a sustainable red abalone fishery and the recovery of sea otters may be able to coexist in southern California. Red abalone populations first must be fully recovered, and sea otter populations would need to be present but controlled. This would ultimately re-create an ecological state that began some 8,000 years ago and persisted until the California Mission period of the nineteenth century.

In January 2010, the Abalone Advisory Group issued its report on the potential for opening a San Miguel Island red abalone fishery. After collecting survey data and performing modeling work, the group decided to continue to refine its model and accumulate additional data, focused on San Miguel Island, over the next 5 to 20 years. The Advisory Group called the modeling work the "best available science" but has yet to explicitly consider the implications of archaeological, paleoecological, and historical data. Potentially missing from its plan is the critical need to monitor red abalone recovery along Santa Rosa and Santa Cruz, in areas where archaeological data suggest abalone become abundantly available when the channel-wide fishery is healthy and thriving. If a commercial (or sport) harvest of red abalone were ever to be reopened along the Northern Channel Islands, one strategy, based on historical ecological perspectives, might be to protect the San Miguel nursery and strictly control and monitor red abalone harvest along the islands to the east. Preserving the nursery population may help prevent the historical cycle of overfishing and serial decline.

Reseeding California's Black Abalone: The Integration of Ancient and Historical Records

By the 1990s, after decades of overfishing and the catastrophic effects of disease, future projections for California black abalone were abysmal. Withering syndrome had decimated populations, especially in the warmer-water, southerly portions of their range, beginning around 1985. Communities on the Channel Islands were especially hard hit, and black abalone virtually disappeared from the rocky intertidal shores of southern California (Miner et al. 2006:108). After two and a half decades of declines, however, there may be a glimmer of hope. Recent intertidal surveys have shown some signs of recovery, especially along localized rocky intertidal shores on the Channel Islands (Butler et al. 2009; Neuman et al. 2010). The jury is still out on the long-term prospects of the species, and scientists and resource managers still have a long road ahead of them.

One potentially effective method for expediting the recovery of black abalone is outplanting (Tegner 2000; California Fish and Game Commission 2005). This technique releases millions of cultured larvae or scores of hatchery-reared juveniles into "optimal habitat" with the hope of creating a thriving population that then can spread to adjacent areas. Since abalone are broadcast spawners, meaning females release unfertilized eggs into the water at the same time that males release sperm, they require densely aggregated adult communities to ensure fertilization and reproduction (Prince et al. 1987). If densities are too low, abalone will not reproduce. This is especially true for black abalone because their larvae have limited dispersal ranges (Chambers et al. 2006; Hamm and Burton 2000; Prince et al. 1988). Outplanting can help establish high enough localized densities of adults so they are able to expand their ranges into new intertidal habitats.

Identifying optimal habitats conducive to larval settlement and early survival can be challenging (e.g., Raimondi 1990), and traditionally, outplanting efforts have focused on areas with bare rock and crustose coralline algae (optimal abalone habitat). Locating these areas is generally accomplished by conducting intertidal surveys and consulting density and distribution data from the several years to decades before modern population crashes. California's Northern Channel Islands, due to their relative isolation, lower levels of human disturbance, and nutrient-rich kelp forests, have been a focus area for these abalone restoration efforts. Outplanting attempts along island shorelines have at times been successful but have failed in equal measure.

One untapped strategy to help improve the success rate is to consult deeper historical records to identify shorelines that have sustained black abalone fishing over millennial timescales. Areas that yielded productive black abalone harvests during the Native American, historical Chinese, and modern commercial fisheries may be relatively resilient to overfishing and long-term depletion. These locations might be the most optimal for outplanting and could act as epicenters of black abalone recovery along island shorelines.

Beginning in 2013, I organized a team of archaeologists and abalone biologists to try to identify these locations using a complex mix of prehistoric, historical, modern, and paleoenvironmental data (Braje et al. 2015). First, we examined spatially discrete historical catch data for southern

California commercial and recreational black abalone fisheries between 1950 and 1993 from the Northern Channel Islands (California Fish and Game 2005). These data provide a pattern of black abalone distribution and relative abundance along the islands, derived from both commercial landings and passenger fishing vessel receipts. The major challenge with the data is that they are divided into relatively coarse-grained blocks (10-×-10-minute, 6600-ha) and offer only low-resolution spatial patterning. They do reveal, however, the prime locations for abalone fishing when sea otters were locally extinct and an intensive abalone fishery had been thriving along the Northern Channel Islands for a century.

These data were compared with deeper time records compiled from prehistoric archaeological shell midden sites and historical Chinese black abalone processing sites and basecamps. Ancient Channel Islanders intensively harvested marine foods, including black abalone and other shellfish, finfish, aquatic birds, and marine mammals, for more than 12,000 years (Erlandson et al. 2009). The intensity of black abalone harvest fluctuated through time, but foraging theory and ethnohistoric records suggest that intertidal black abalone were an easily accessible and highly ranked resource (see Braje, Kennett et al. 2007) throughout the Holocene. When black abalone were available, humans would have targeted them, consumed the meat, and deposited the shells in ancient trash piles, near locations where they were collected. These behaviors make the shells' relative abundances in archaeological assemblages an excellent proxy for prehistoric availability along adjacent coastlines. Though most of the Northern Channel Island shorelines have been surveyed for archaeological sites, the number of shell middens that have been radiocarbon dated and quantified is uneven, with less research on the southern shore of Santa Rosa Island and the eastern and southern coastlines of Santa Cruz Island. Despite these gaps, the available archaeological data are a robust measure of trans-Holocene black abalone fishing along the northern islands.

When these data are combined with the distribution of historical Chinese black abalone foraging camps, a nearly 10,000-year record of black abalone abundances and distributions is revealed. This record spans fluctuations in ancient human population densities, natural climatic shifts, and habitat alterations. The combination of datasets identifies locations where black abalone populations have thrived over deep time across millennia and general areas where outplanting efforts might have the greatest potential for success.

Nearly all of the thousands of well-preserved Northern Channel Island prehistoric shell middens contain at least some black abalone shell. After California mussel, black abalone is one of the most ubiquitous marine invertebrates spanning the last 10,000 years. Following methodological benchmarks established by earlier studies on the historical ecology of red abalone fishing (see Braje et al. 2009), we compiled data from all excavated archaeological sites that contain at least 5 percent black abalone shell (represented by total shell weight). These locations then were compared against the record of Chinese black abalone fishing and the recent record of commercial black abalone fishing. The major challenge in this effort stems from the incongruence between ancient, historical, and modern datasets. Ancient and historical archaeological data are based on raw weights of black abalone shells transported to terrestrial refuse piles, and modern commercial data are catch records of whole abalone (often shells and meat) transported to the mainland. Even though these data are collected and reported in different ways, they provide a relative estimate of black abalone abundance through time, which may be analyzed to identify overall patterning.

It is also important, however, to confront the methodological limitations of these historical ecological data. Archaeological, historical, and contemporary catch data represent harvest or landing records, which do not always accurately reflect species abundance in the wild. Catch data can be influenced by many factors, such as harvest effort, technological innovation, management policies, access, travel costs, and resource switching. Variations in these dynamics can result in the comparison of disparate data and create a level of uncertainty over what landings data (contemporary, historical, and archaeological) actually represent. High landings, for example, do not always signal a healthy, thriving population; they may instead reflect increased harvest effort of a declining population (see the red abalone example detailed earlier or the historical example of Atlantic cod; Kurlansky 1997). Such comparisons over deep time, however, offer an effective first step in identifying habitats and coastlines where species have thrived over thousands of years, through changing climatic conditions, varied levels of human and other animal predation, and anthropogenic changes. Once identified, inherent assumptions can be ground-truthed with site-specific ecological assessments.

The highest densities of black abalone across the prehistoric, historical, and modern fisheries are found along northwestern, northeastern, and

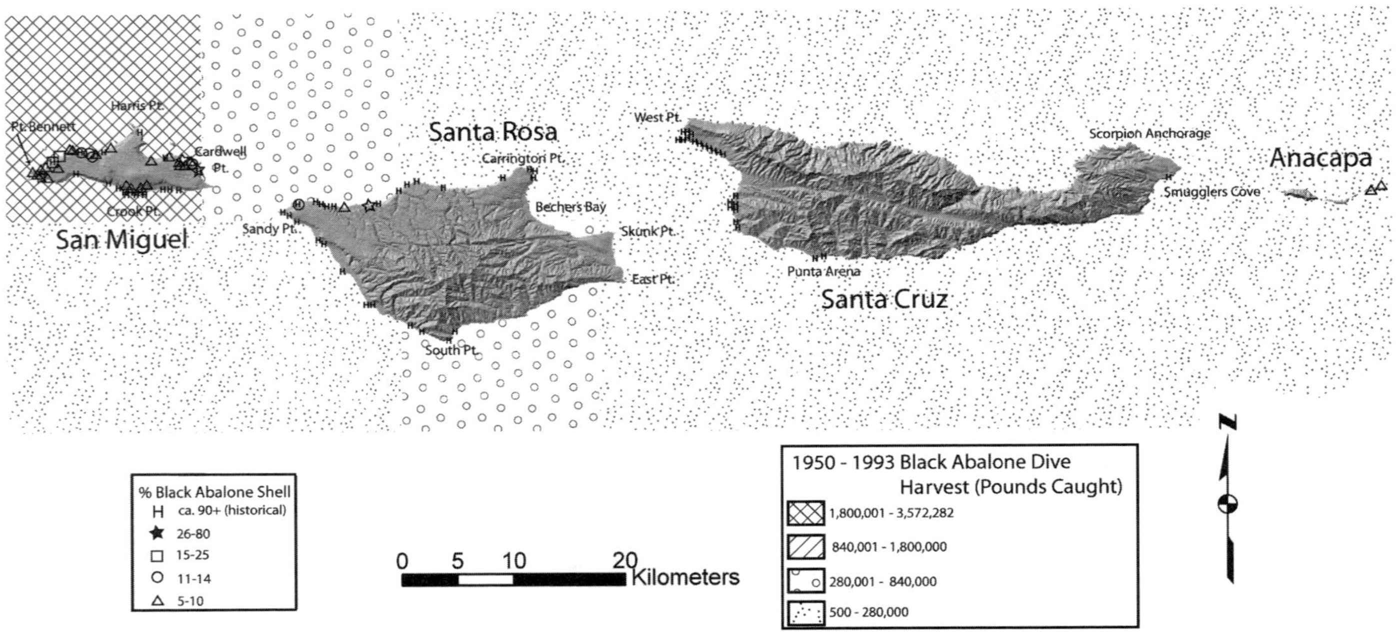

FIGURE 8.5. Map of Northern Channel Islands historical commercial black abalone fishing areas from 1950 to 1993 overlaid with archaeological shell midden sites with greater than 5 percent black abalone shell and with historical Chinese abalone fishing sites. For conversion to SI units, 1 pound ≈ 0.45 kg.

south-central San Miguel Island (Figure 8.5). Coastline stretching from the western flank of Harris Point to the Point Bennett area contains 16 prehistoric shell midden site components, ranging in age from 9700 to 350 cal BP, with black abalone densities between 28.9 and 5.1 percent, along with four historical black abalone middens (Table 8.1). Shorelines between the eastern flank of Harris Point to Cardwell Point produced 10 prehistoric shell midden components ranging in age from 9500 to 3300 cal BP with black abalone densities between 30.6 and 6.1 percent, along with two historical black abalone middens. The Crook Point area produced four prehistoric shell midden sites ranging in age from 9500 to 1200 cal BP with black abalone densities between 10.0 and 5.6 percent, along with seven historical black abalone middens. All of these San Miguel Island areas fall within the highest-ranked block for modern commercial black abalone fishing. These locations supported long-term, intensive black abalone fisheries throughout the Holocene, suggesting that they may contain the right mix of environmental conditions for black abalone outplanting and recovery. Black abalone populations at these locations span warm- and cold-water intervals and are not tied to natural climatic fluctuations: half the high-density samples occurred during warm-water periods and half during cold-water episodes (Figure 8.6).

Combining prehistoric and historical archaeological and modern ecological data that identify locations with favorable environmental factors, current patterns, and recruitment for black abalone can provide compelling evidence for prime restoration locations. The trans-Holocene record of black abalone fishing along the Channel Islands highlights areas that were long-term "hot spots" for black abalone recruitment and reproduction before the historical extirpation of sea otters and the rise of intensive modern commercial fisheries. Since few specific recommendations were made for the best restoration locations in California's Abalone Recovery and Management Plan, these areas may provide the best sites for restoration work and outplanting. Archaeological records are especially important for downscaling to specific black abalone restoration sites within large-scale regions identified using commercial fishing records. Modern intertidal surveys in these areas, adjacent to archaeological shell middens (prehistoric or historical) with high proportions of black abalone shell, may be especially fruitful for identifying black abalone outplanting habitats.

TABLE 8.1. Radiocarbon Age and Percentage of Black Abalone Shell from Archaeological Shell Middens on the Northern Channel Islands with Quantified Faunal Data.

Site #	Provenience	Age (cal BP)[a]	% Black Abalone	Reference
San Miguel Island				
522	Unit 1	9700	9.5	Erlandson et al. 2005
608	Bulk samples 1 & 2	9500	10.0	Braje et al. 2012
548	Unit 1	9500	5.8	Erlandson et al. 2005
261	Col. E-6, Str. E4	9100	6.2	Braje et al. 2012
507	Unit 1	9000	17.0	Braje et al. 2012
261	Col. E-6, Str. E3	9000	30.6	Erlandson et al. 2005
261	Col. E-6, Str. E2	8900	9.9	Erlandson et al. 2005
606	Units 1 & 2	8800	7.4	Braje et al. 2012
608W	Unit 1	8600	5.6	Garcia 2007
603	Bulk sample, Str. 7	7900	12.1	Ainis et al. 2011
527	Bulk sample 1	5100	7.8	Braje et al. 2012
396	Bulk samples 1 & 2	4800	6.4	Braje et al. 2012
161	Unit 1	4500	10.7	Rick unpublished notes
603	Bulk sample, Str. 5	4370	10.9	Ainis et al. 2011
603	Str. 4	4320	6.5	Braje et al. 2012
628	Col. 1, L. 2–3	4200	6.8	Braje et al. 2012
603	Bulk sample, Str. 3	4060	28.1	Ainis et al. 2011
503N	Str. 2	3650	16.4	Erlandson et al. 2005
261	Col. E-6, Str. A	3300	6.0	Erlandson et al. 2005
525	Str. 3	2800	28.9	Erlandson et al. 2005
525D	Str. 9	1700	9.5	Erlandson et al. 2005
510	N Profile, Str. 6	1200	5.6	Braje et al. 2012
232	Col. 1 & 2	1200	10.0	Braje et al. 2012
468	Unit 2, Str. III	1200	11.5	Braje et al. 2012
481	Column sample 1	1200	5.1	Braje et al. 2012
481	Unit 1a, Str. 1	1170	6.7	Braje et al. 2012
468	Unit 2, Str. III	1100	11.5	Braje et al. 2012
468	Unit 1	850	5.6	Braje et al. 2012
525	D Profile, Str. 3	550	9.4	Braje et al. 2012
470	Str. 1	350	13.3	Erlandson et al. 2005
Santa Rosa Island				
6	Bulk sample	9300	43.4	Braje et al. 2012
15	Unit 2, 40–50 cmbs.	590	12.1	Jazwa et al. 2012
2	Unit 1	130	5.5	Braje et al. 2012
Anacapa Island				
2	Bulk sample	2900	> 5	Rick unpublished notes
3	Bulk sample	3200	6.2	Rick unpublished notes

[a] Site age is based on the approximate midpoint of one or more calibrated radiocarbon dates.

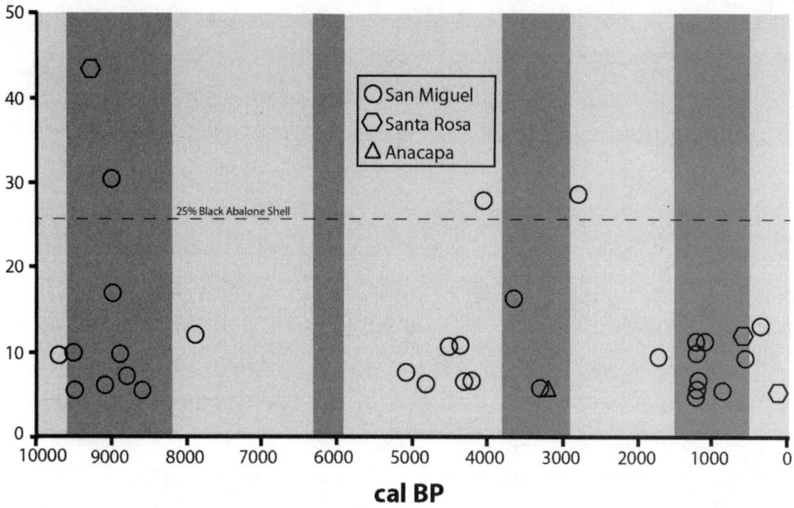

FIGURE 8.6. Percentage of black abalone (*y*-axis) from 35 shell midden sites or site components on the Channel Islands compared with Holocene Santa Barbara Basin sea surface temperature (SST). Inferred SST is based on δ18O of *Globigerina bulloides* (surface-dwelling species of foraminifera) from varved sediments in the Santa Barbara Basin (see Kennett 2005:64–69). Light gray shading along the *x*-axis represents warm-water cycles and dark gray represents cold-water cycles through the Holocene, based on a 12.5°C Holocene average. Discontinuities between SST and black abalone abundance suggest that black abalone abundance was not solely controlled by SST.

The important lesson to be learned is that the combination of modern, historical, and ancient data, spanning hundreds to thousands of years, can help maximize the potential of restoration efforts and expedite the recovery and expansion of black abalone communities in southern California. No longer can scientists, resource managers, and conservationists afford to work alone. By consulting historical and archaeological records, we can better understand how, where, and why shellfish species have been resilient to heavy predation, natural climate change, commercial exploitation, and disease over long time periods.

The Future Past of Global Seascapes and California Abalone Fisheries

After 12,000 years of nearly continuous human exploitation and thriving populations, California abalone face an uncertain future. For some species such as whites, it may be only a matter of time before wild populations are just a distant memory. For others, localized rebounds offer a reason for optimism; we may have a second chance to build healthy, effective, and sustainable management plans. Success may rest, in no small part, on our willingness to heed the lessons of history and integrate new data, establish new disciplines, and develop new methodologies.

The collapses of commercial fisheries around the globe are well known to historians and ecologists alike. It is no secret that, since 1950, 80 percent of the world's industrial fisheries have crashed or are over- or fully exploited. Apex predators (and the humans who eat them) such as tuna, salmon, and swordfish are increasingly full of mercury, PCBs, and other toxins. Unthinkably large concentrations of trash, chemical sludge, and plastic circulate our world's oceans in five massive gyres. Overfishing, pollution, exploding human populations, rising global temperatures, disease, ocean acidification, and myriad other challenges stress resources and have pushed marine systems to their limits. What are left are entire ecosystems threatened with extinction—coral reefs bleached and diseased, hundreds of marine dead zones teeming with hypoxic waters, high-trophic-level-fish reduced to just a fraction of their historical abundances.

The shifting-baselines concept has been revolutionary in helping us rethink how to address these crises and meet the enormous challenges we now face. Historical perspectives can help us evaluate the perilous state of our marine ecosystems and fisheries and determine an accurate measure of declines. Historical ecologists can track long-term changes in species abundance and distribution, biogeography, habitat, food webs, and much more. Despite these advantages, many resource managers and scientists have been slow to accept or integrate the perspectives of historical scientists. Some have yet to recognize the utility of such perspectives, but most struggle with practical ways to integrate these data. This is not surprising since historical ecology is still an evolving field, striving to develop a set of methodological contributions. The most recalcitrant hurdle in historical ecology's push for widespread acceptance may be the challenges faced when disparate datasets, often reported and collected in very different ways, are combined across disciplines.

The underlying problem is that historical ecologists must sacrifice precision for generality and realism (Jackson and Alexander 2011:4). Take, for example, general patterns revealed by investigating the deep history of California red abalone fishing. Archaeological data reveal that red abalone harvesting was a productive fishing strategy for over 12,000 years on the

Northern Channel Islands. Average prey sizes were relatively small during this ancient fishery (see Erlandson, Rick, Braje et al. 2008), but despite heavy predation pressure, there is no evidence for localized extirpation or long-term resource stress. For 12,000 years, San Miguel Island waters offered extremely productive, continuously exploited red abalone fishing grounds. During favorable climatic conditions, red abalone became more abundantly available over larger portions of the Santa Barbara Channel. This general temporal and spatial patterning cannot be tested, in any traditional sense, against modern survey data. Raw counts of red abalone density along San Miguel Island, for example, will never be available over this 12,000-year history. To integrate these findings into modern management, managers must be comfortable with exchanging a lack of precision over short time scales for generalized patterning over deep time scales. Such a strategy represents a sea change in how fisheries science and restoration biology have long been accomplished.

In some cases, the application of ancient and historical data is straightforward. Compare, for example, the ancient fishery with the historical and modern commercial fishery, and clearly, the spectacularly productive black abalone fishery of the mid- to late nineteenth century and the red abalone fishery of the early to mid-twentieth century were ecological aberrations. The extraordinary sizes and densities of abalone were unprecedented over the last 10,000 years (see Erlandson, Rick, Braje et al. 2008), the result of sea otter overhunting and the release of abalone from predation pressure. The yields by Chinese intertidal fishermen for black abalone and Japanese hardhat divers for red abalone have created a set of unrealistic expectations for California abalone populations and the productivity of the fishery. The dramatic collapse of commercial (and sport) abalone fishing in California demonstrates that we cannot expect to feed a global abalone meat and shell market. Instead, our goal should be to rebuild and rebalance ecosystems, to reestablish wild abalone communities, and to repair degraded kelp forest ecosystems.

Deep temporal perspectives become more challenging to incorporate, however, when arguments include reintroducing apex sea otters as keystone ecosystem managers. For many, the goal of restoring abalone fisheries in California should be to re-create the historical fishery. Many fishermen and managers argue that sea otters cannot coexist in California alongside a commercial abalone fishery—the competition would be unsustainable. Such a view does not consider the millennia-long coexistence of a Native American abalone fishery and sea otters. The challenge, of course, will be to find the balance between self-sustaining abalone communities, regulated sea otter populations, and human fisheries. Our ultimate goal should be to rebuild localized and sustainable harvests, structured by employing strategies learned from the 12,000-year-long Native American harvest.

The unfortunate reality is that most marine ecologists, fisheries biologists, resource managers, and policy makers rely on quantitative estimates of population size through time rather than functional processes, even if they recognize the potential contributions of historical perspectives. Historical ecologists will never be able to count the number of abalone along an ancient watershed. Historical data are ill equipped to offer the same level of precision as modern data. It is easier, then, to focus on recent datasets and track changes over the last several decades, all the while ignoring the potential lessons of deep history.

In many ways, this divide boils down to basic epistemological questions. Environmental management and conservation laws in the United States require that the "best available science (BAS)" is applied when any policy decision is made (Sullivan et al. 2006). Accumulating knowledge to address a management issue may not be contentious or particularly difficult. Assigning value to the quality of that information and determining what constitutes the BAS, however, can create conflicts among scientists, managers, and stakeholders. The increasingly interdisciplinary nature of science and the rise of historical ecology challenge our basic assumptions of what constitutes high-quality science. The only solution might be to strive for enhanced understandings of the potential contributions of all role players (see Clark and Majone 1985). This would mean designing projects, from the ground up, that include ecologists, biologists, archaeologists, resource managers, policy makers, stakeholders, and other interested scientists. Such a strategy creates dialogue across disciplines and improves our ability to understand the potential contributions of unique datasets. No longer can we afford to develop policy based on partial perspectives. Rather, we need to build interdisciplinary teams to address scientific and management questions from a common framework. In doing so, we can generate critical discussion and find ways "each partial perspective can complement and reinforce the others" (Clark and Majone 1985:17)

Despite all the ambiguity we face and the immense challenges on our horizon, I am certain of one thing: addressing today's complex environmental issues will take a community of dedicated scientists, managers, and concerned citizens. Solutions will come as interdisciplinary teams rally to solve complicated issues for which they can offer unique perspectives. Armed with new and creative solutions, we can restore abalone stocks in California and repair the damage caused by decades of overfishing, pollution, disease, ecological degradation, and mismanagement. Conservative management approaches, informed by the lessons of history, can result in a renewed and sustainable fishery. Of course, none of this will be possible if we don't begin while we still have a chance.

References

Ainis, Amira F., René L. Vellanoweth, Troy W. Davis, Jon M. Erlandson, and Torben C. Rick

2011 Changes in Marine Subsistence on San Miguel Island from 8,500 to 2,400 Years Ago: Analysis of Bulk Samples from Cave of the Chimney (CA-SMI-603). *Journal of California and Great Basin Anthropology* 31(1):59–79.

Allan, Sarah (editor)

2002 *The Formation of Chinese Civilization: An Archaeological Perspective*. Yale University Press, New Haven, Connecticut.

Ambrose, Stephen E.

2001 *Nothing Like It in the World: The Men Who Built the Transcontinental Railroad, 1863–1869*. Touchstone, New York.

Anderson, Andrew Wallace, and Walter H. Stolting

1953 *Survey of the Domestic Tuna Industry*. U.S. Department of the Interior, Fish and Wildlife Service, Washington, D.C.

Anonymous

1881 Abalone Jewelry and Its Source. *Harper's Bazaar* 14:38–39.

Armentrout-Ma, L. Eve

1981 Chinese in California's Fishing Industry, 1850–1941. *California History* 60(2):142–157.

Associated Press

2011 Giant Tuna Sells for Record $396,000 in Tokyo. http://www.cbsnews.com/news/giant-tuna-sells-for-record-396000-in-tokyo/.

Axford, Michael

1984 Four Years of Archaeological Investigations on San Clemente Island, California. Ms. on file, Natural Resource Office, North Island Naval Reservation, San Diego, California.

1987 Late Historic Chinese Abalone Collectors on San Clemente Island. Ms. on file, Natural Resource Office, North Island Naval Reservation, San Diego, California.

Bain, David Haward

1999 *Empire Express: Building the First Transcontinental Railroad*. Penguin Books, New York.

Balée, William, and Clark L. Erickson (editors)
2006 *Time and Complexity in Historical Ecology: Studies in the Neotropical Lowlands*. Columbia University Press, New York.
Bancroft, Hubert H.
1890 *The Works of Hubert Howe Bancroft: Vol. XXIV, History of California; Vol. VII, 1860–1890*. History Company, San Francisco.
Barfield, Thomas J.
1989 *The Perilous Frontier: Nomadic Empires and China, 221 BC to AD 1757*. Blackwell, Cambridge, Massachusetts.
Barlow, Jeffrey, and Christine Richardson
1979 *China Doctor of John Day*. Binford and Mort, Portland, Oregon.
Barnes, Frank C.
1965 *Cartridges of the World*. DBI Books, Northfield, Illinois.
Baxter, R. Scott
2008 The Response of California's Chinese Populations to the Anti-Chinese Movement. *Historical Archaeology* 42(3):29–36.
Baxter, R. Scott, and Rebecca Allen
2002 Archaeological Investigations of Life within the Woolen Mills, Chinatown, San Jose. In *The Chinese in America: A History from Gold Mountain to the New Millennium*, edited by S.L. Cassel, pp. 381–398. Altamira Press, Walnut Creek, California.
Bentz, Linda
1996 Chinese Abalone Collectors of Santa Rosa Island. Unpublished report presented to Channel Islands National Park, Ventura, California.
Bentz, Linda, and Robert Schwemmer
2002 The Rise and Fall of the Chinese Fisheries in California. In *The Chinese in America: A History from Gold Mountain to the New Millennium*, edited by S.L. Cassel, pp. 140–155. Altamira Press, Walnut Creek, California.
Berryman, Judy Ann
1995 Archival Information, Abalone Shell, Broken Pots, Hearths, and Windbreaks: Clues to Identifying Nineteenth Century California Abalone Collection and Processing Sites, San Clemente Island: A Case Study. Unpublished Ph.D. dissertation, University of California, Riverside.
Bloeser, Jennifer Anne
1999 *Diminishing Returns: The Status of West Coast Rockfish*. Pacific Marine Conservation Council, Astoria, Oregon.
Bodkin, James
2000 Sea Otters. *Alaska Geographic* 27(2):74–92.
Bolster, W. Jeffrey
2012 *The Mortal Sea: Fishing the Atlantic in the Age of Sail*. Belknap Press, Cambridge, Massachusetts.
Bonnot, Paul
1930 Abalones in California. *California Fish and Game* 16:15–23.

Borthwick, J.D.
1857 *Three Years in California*. William Blackwood and Sons, Edinburgh, Scotland.

Braje, Todd J.
2007 Geographic and Temporal Variability of Middle Holocene Red Abalone Middens on San Miguel Island, California. *Journal of California and Great Basin Anthropology* 27(2):153–164.
2010 *Modern Oceans, Ancient Sites: Archaeology and Marine Conservation on San Miguel Island, California*. University of Utah Press, Salt Lake City.

Braje, Todd J., and Linda Bentz
2015 The Archaeology of Overseas Chinese Abalone Fishermen on Santa Rosa Island, California. *Journal of Chinese Overseas* 11:87–103.

Braje, Todd J., Julia G. Costello, Jon M. Erlandson, and Robert DeLong
2014 Of Seals, Sea Lions, and Abalone: The Archaeology of a Historic Multi-ethnic Basecamp on San Miguel Island, California. *Historical Archaeology* 48(2):122–142.

Braje, Todd J., and Jon M. Erlandson
2006 Historical Abalone Fishers of San Miguel Island, California. In *Proceedings of the Society for California Archaeology, Volume 19*, edited by S.A. Waechter, D. Laylander, and G.G. White, pp. 21–26. SCA Business Office, Chico, California.
2007 Measuring Subsistence Specialization: Comparing Historic and Prehistoric Abalone Middens on San Miguel Island, California. *Journal of Anthropological Archaeology* 26:474–485.
2008 Early Maritime Technology from Western San Miguel Island, California. *Current Research in the Pleistocene* 25:31–32.

Braje, Todd J., Jon M. Erlandson, and Torben C. Rick
2007 An Historic Chinese Abalone Fishery on California's Northern Channel Islands. *Historical Archaeology* 41(4):117–128.

Braje, Todd J., Jon M. Erlandson, Torben C. Rick, Paul K. Dayton, and Marco B.A. Hatch
2009 Fishing from Past to Present: Continuity and Resilience of Red Abalone Fisheries on the Channel Islands, California. *Ecological Applications* 19(4):906–919.

Braje, Todd J., Douglas J. Kennett, Jon M. Erlandson, and Brendan J. Culleton
2007 Human Impacts on Nearshore Shellfish Taxa: A 7,000 Year Record from Santa Rosa Island, California. *American Antiquity* 72(4):735–756.

Braje, Todd J., and Torben C. Rick
2013 From Forest Fires to Fisheries Management: Anthropology, Conservation Biology, and Historical Ecology. *Evolutionary Anthropology* 22: 303–311.
2015 Identifying Specialized Nineteenth Century Fishing Camps on the Northern Channel Islands: An Application of AMS Radiocarbon Dating. *Radiocarbon* 57(5):909–916.

Braje, Todd J., and Torben C. Rick (editors)
2011 *Human Impacts on Seals, Sea Lions, and Sea Otters: Integrating Archaeology and Ecology in the Northeast Pacific*. University of California Press, Berkeley.

Braje, Todd J., Torben C. Rick, and Jon M. Erlandson
2011 Rockfish in the Longview: Applied Archaeology and Conservation of Pacific Red Snapper (Genus *Sebastes*) in Southern California. In *Zooarchaeologial Perspectives on Issues in Conservation Biology*, edited by S. Wolverton, R.L. Lyman, and C.R. Randklev, pp. 157–178. University of Arizona Press, Tucson.
2012 A Trans-Holocene Historical Ecological Record of Shellfish Harvesting on California's Northern Channel Islands. *Quaternary International* 264(20):109–120.

Braje, Todd J., Torben C. Rick, Jon M. Erlandson, Laura Rogers-Bennett, and Cynthia Catton
2015 Long-Term Ecological History Can Inform Restoration Site Selection: The Case of Black Abalone (*Haliotis cracherodii*) along California's Channel Islands. *Aquatic Conservation*. doi: 10.1002/aqc.2561.

Butler, John, Andrew DeVogelaere, Rick Gustafson, Chris Mobley, Melissa Neuman, Dan Richards, Scott Rumsey, Barbara Taylor, and Glenn VanBlaricom
2009 Status Review Report for Black Abalone. National Marine Fisheries Service, Long Beach, California.

Caire, Helen
1993 *Santa Cruz Island: A History and Recollections of an Old California Rancho*. Arthur H. Clark, Spokane, Washington.

California Fish and Game Commission
1910 *Twenty-First Biennial Report of the Board of Fish and Game Commissioners of the State of California*. Fish and Game Commission, Sacramento, California.
2001 *California's Living Marine Resources: A Status Report*. California Department of Fish and Game, Marine Region, Sacramento.
2005 Abalone Recovery and Management Plan. California Department of Fish and Game, Marine Region, Sacramento.

Campbell, Persia Crawford
1923 *Chinese Coolie Emigration to Countries within the British Empire*. P.S. King and Son, London.

Canton, John Dean
1891 Abalone and Squid. *Forest and Stream* 36:474–475.

Carlton, James T.
1998 Apostrophe to the Ocean. *Conservation Biology* 12(6):1165–1167.

Carter, Nancy Carol
2007 San Diego Olives: Origins of a California Industry. *Journal of San Diego History* 54(3):137–161.

Chambers, Melinda D., Glenn R. VanBlaricom, Lorenz Hauser, Fred Utter, and Carolyn S. Friedman
2006 Genetic Structure of Black Abalone (*Haliotis cracherodii*) Populations in the California Islands and Central California Coast: Impacts of Larval Dispersal and Decimation from Withering Syndrome. *Journal of Experimental Marine Biology and Ecology* 331(2):173–185.
Chan, Sucheng
1986 *This Bittersweet Soil: The Chinese in California Agriculture, 1860–1910*. University of California Press, Berkeley.
1991 *Asian Americans: An Interpretive History*. Twayne, New York.
Chang, Iris
2003 *The Chinese in America*. Penguin Books, New York.
Chang, Kwang-Chih
1986 *The Archaeology of Ancient China*. Yale University Press, New Haven, Connecticut.
Chen, Jack
1981 *The Chinese of America*. Harper and Row, San Francisco.
Chiles, Frederic Caire
2011 *Justinian Caire and Santa Cruz Island: The Rise and Fall of a California Dynasty*. University of Oklahoma Press, Norman.
Chinn, Thomas W., Mark H. Lai, and Phillip P. Choy (editors)
1969 *A History of the Chinese in California: A Syllabus*. Chinese Historical Society of America, San Francisco.
Chung, Sue Fawn
2011 *In Pursuit of Gold: Chinese American Miners and Merchants in the American West*. University of Illinois Press, Urbana.
Cicin-Sain, Biliana, John E. Moore, and Alan J. Wyner
1977 *Management Approaches for Marine Fisheries: The Case of California Abalone*. Institute of Marine Resources, University of California, La Jolla.
Clark, William C., and Giandomenico Majone
1985 The Critical Appraisal of Scientific Inquiries with Policy Implications. *Science, Technology, and Human Values* 10(3):6–19.
Collins, Donna
1987 Emerging Archaeological Evidence of the Chinese Market Fisheries of Early California. *Pacific Coast Archaeological Society Quarterly* 23(2): 63–68.
Collins, J.W.
1888 *Report on the Fisheries of the Pacific Coast of the United States*. Report to the United States Commission of Fish and Fisheries. Washington, D.C.
Coolidge, Mary Roberts
1909 *Chinese Immigration*. Henry Holt, New York.
Costello, Julia G., Kevin Hallaran, and Keith Warren
2004 The Luck of Third Street: Historical Archaeology Data Recovery Report

for the Caltrans District 8 San Bernardino Headquarters Demolition Project. Prepared by Foothill Resources, Ltd., and Applied Earthworks, Inc., for California Department of Transportation, District 8, San Bernardino, California.

Costello, Julia G., Kevin Hallaran, Keith Warren, and Margie Akin

2008 The Luck of Third Street: Archaeology of Chinatown, San Bernardino, California. *Historical Archaeology* 42(3): 136–151.

Costello, Julia G., and Mary L. Maniery

1988 *A Catalogue of Artifacts Recovered from the 1915 Asian Community of Walnut Grove, California*. Occasional Paper No. 16. Institute of Archaeology, University of California, Los Angeles.

Costello, Julia G., Adrian Praetzellis, Mary Praetzellis, Erica S. Gibson, Judith Marvin, Michael D. Meyer, Grace H. Ziesing, Susan K. Goldberg, Sherri M. Gust, Madeline Hirn, William Marvin Mason, Elaine-Maryse Solari, and Suzanne B. Stewart

1999 Historical Archaeology at the Headquarters Facility Project Site, the Metropolitan Water District of Southern California. Prepared by Foothill Resources, Ltd., Applied Earthworks, Inc., and the Anthropological Studies Center, Sonoma State University Academic for the Metropolitan Water District of Southern California, Los Angeles.

Costello, Julia, Jane Russell, and Susan Stratton

1998 CA-CAL-1564/H, Locus 4: Selkirk Ranch Homestead with Chinese Hearth and Forge; Excavation Report. Prepared by Foothill Resources, Ltd., Mokelumne Hill, California. Report on file at the Central California Information Center, Turlock.

Courchamp, Franck, Elena Angulo, Philippe Rivalan, Richard J. Hall, Laetitia Signoret, Leigh Bull, and Yves Meinard

2006 Rarity Value and Species Extinction: The Anthropogenic Allee Effect. *PLoS Biology* 4(12):e415.

Courtwright, David T.

2001 *Dark Paradise: A History of Opium Addiction in America*. Harvard University Press, Cambridge, Massachusetts.

Cox, Keith W.

1962 *Review of the Abalone of California, Family Haliotidae*. Fish Bulletin No. 118. California Department of Fish and Game, Sacramento.

Cranmer-Byng, J.L. (editor)

1963 *An Embassy to China: Being the Journal Kept by Lord Macartney during His Embassy to the Emperor Ch'ien-lung, 1793–1794*. Archon Books, Hamden, Connecticut.

Crawford, Gary W.

2011 Early Rice Exploitation in the Lower Yangzi Valley: What Are We Missing? *The Holocene* 22(6):613–621.

Crawford, Gary W., and Chen Shen

1998 The Origins of Rice Agriculture: Recent Progress in East Asia. *Antiquity* 72(278):858–866.

Crawford, Mark
1999 *Encyclopedia of the Mexican-American War*. ABC-CLIO, Santa Barbara, California.

Crawford, Richard
2009 San Diego Had Its Own Cannery Row. *San Diego Union-Tribune*, June 18. http://www.sandiegouniontribune.com/news/2009/jun/18/1mc18history 19655-san-diego-had-its-own-cannery-ro/.

Croker, R.S.
1931 Abalones. In *The Commercial Fish Catch of California for the Year 1929*, pp. 58–72. Fish Bulletin No. 30. California Department of Fish and Game, Sacramento.

Crumley, Carole L. (editor)
1994 *Historical Ecology: Cultural Knowledge and Changing Landscapes*. School of American Research Press, Santa Fe, New Mexico.

Daniels, Roger
1988 *Asian America: Chinese and Japanese in the United States since 1850*. University of Washington Press, Seattle.

Dartt-Newton, Deana, and Jon M. Erlandson
2006 Little Choice for the Chumash: Colonialism, Cattle, and Coercion in Mission Period California. *American Indian Quarterly* 30(3–4):416–430.

Dary, David
1995 *Seeking Pleasure in the Old West*. Alfred A. Knopf, New York.

Davis, Gary E., Peter L. Haaker, and Daniel V. Richards
1998 The Perilous Condition of White Abalone, *Haliotis sorenseni* Bartsch, 1940. *Journal of Shellfish Research* 17(3):871–875.

Dayton, Paul K., Mia J. Tegner, Peter B. Edwards, and Kristin L. Riser
1998 Sliding Baselines, Ghosts, and Reduced Expectations in Kelp Forest Communities. *Ecological Applications* 8(2):309–322.

Denevan, William M. (editor)
1992 *The Native Population of the Americas in 1492*. 2nd ed.. University of Wisconsin Press, Madison.

Diehl, Michael, Jennifer A. Waters, and J. Homer Thiel
1998 Acculturation and the Composition of the Diet of Tucson's Overseas Chinese Gardeners at the Turn of the Century. *Historical Archaeology* 32(4):19–33.

Douros, William J.
1985 Density, Growth, Reproduction, and Recruitment in an Intertidal Abalone: Effects of Intraspecific Competition and Prehistoric Predation. Unpublished Master's thesis, University of California, Santa Barbara.

Downey, Lynn
2007 *Levi Strauss & Co.* Images of America. Arcadia Publishing, Charleston, South Carolina.

Dreyer, Edward
2007 *Zheng He: China and the Oceans in the Early Ming Dynasty, 1405–1433*. Pearson Longman, New York.

Dugan, Jennifer E., and Gary E. Davis
1993 Applications of Marine Refugia to Coastal Fisheries Management. *Journal of Fisheries and Aquatic Sciences* 50(9):2029–2042.
Duyvendak, J.J.L.
1939 The True Dates of the Chinese Maritime Expeditions in the Early Fifteenth Century. *T'oung Pao* 34(5):341–413.
Eames, David B.
1995 *San Francisco Street Secrets: The Stories behind San Francisco Street Names.* Gem Guides, Baldwin Park, California.
Eames, Ninetta
1892 Three Weeks on a Weird Island. *Frank Leslie's Popular Monthly* XXXIV (6–42):657–671.
Earle, Homer P.
1896 The Santa Barbara Islands. *Land of Sunshine* 5:227–230.
Edholm, C.L.
1913 Steaks and Pearls from the Abalone. *Overland Monthly* 62(4):383–386.
Edwards, Charles Lincoln
1913 *The Abalone Industry in California.* Fish Bulletin No. 1. State of California Fish and Game Commission, Sacramento.
Ellis, Richard
2003 *The Empty Ocean.* Island Press/Shearwater Books, Washington, D.C.
Elvin, Mark
1993 Three Thousand Years of Unsustainable Growth: China's Environment from Archaic Times to the Present. *East Asian History* 6:7–46.
Erlandson, Jon M., and Kevin Bartoy
1995 Cabrillo, the Chumash, and Old World Diseases. *Journal of California and Great Basin Anthropology* 17(2):153–173.
Erlandson, Jon, Robert Losey, and Neil Peterson
2001 Early Maritime Contact on the Northern Oregon Coast: Some Notes on the 17th Century Nehalem Beeswax Ship. In *Changing Landscapes: "Telling Our Stories"; Proceedings of the Fourth Annual Coquille Cultural Preservation Conference, 2000*, edited by J. Younker, M.A. Tveskov, and D.G. Lewis, pp. 45–53. Coquille Indian Tribe, North Bend, Oregon.
Erlandson, Jon M., Torben C. Rick, and Todd J. Braje
2009 Fishing Up the Food Web? 12,000 Years of Maritime Subsistence and Adaptive Adjustments on California's Channel Islands. *Pacific Science* 63(4):711–724.
Erlandson, Jon M., Torben C. Rick, Todd J. Braje, Molly Casperson, Brendan Culleton, Brian Fulfrost, Tracy Garcia, Daniel A. Guthrie, Nicholas Jew, Douglas J. Kennett, Madonna L. Moss, Leslie Reeder, Craig Skinner, Jack Watts, and Lauren Willis
2011 Paleoindian Seafaring, Maritime Technologies, and Coastal Foraging on California's Channel Islands. *Science* 331(6021):1181–1185.

Erlandson, Jon M., Torben C. Rick, Todd J. Braje, Alexis Steinberg, and René L. Vellanoweth
2008 Human Impacts on Ancient Shellfish: A 10,000 Year Record from San Miguel Island. *Journal of Archaeological Science* 35:2144–2152.
Erlandson, Jon M., Torben C. Rick, James A. Estes, Michael H. Graham, Todd J. Braje, and René Vellanoweth
2005 Sea Otters, Shellfish, and Humans: 10,000 Years of Ecological Interaction on San Miguel Island, California. In *Proceedings of the Sixth California Islands Symposium*, edited by D.K. Garcelon and C.A. Schwemm, pp. 9–21. Institute for Wildlife Studies, Arcata, California.
Erlandson, Jon M., Torben C. Rick, Douglas J. Kennett, and Phillip L. Walker
2001 Dates, Demography, and Disease: Cultural Contacts and Possible Evidence for Old World Epidemics among the Protohistoric Island Chumash. *Pacific Coast Archaeological Society Quarterly* 37(3):11–26.
Erlandson, Jon M., Torben C. Rick, and René L. Vellanoweth
2008 *A Canyon through Time: Archaeology, History, and Ecology of the Tecolote Canyon Area, Santa Barbara, California*. University of Utah Press, Salt Lake City.
Estes, Don
1977 Kondo Masaharu and the Best of All Fishermen. *Journal of San Diego History* 23(3). http://www.sandiegohistory.org/journal/77summer/kondo.htm
Fairbank, John King
1986 *The Great Chinese Revolution, 1800–1985*. Harper and Row, New York.
Feder, Kenneth L.
2008 *Frauds, Myths, and Mysteries: Science and Pseudoscience in Archaeology*. 7th ed. McGraw-Hill, New York.
2011 *The Past in Perspective: An Introduction to Human Prehistory*. Oxford University Press, New York.
Feeny, David, Susan Hanna, and Arthur F. McEvoy
1996 Questioning the Assumptions of the "Tragedy of the Commons" Model of Fisheries. *Land Economics* 72(2):187–205.
Fike, Richard E.
1987 *The Bottle Book: A Comprehensive Guide to Historic, Embossed Medicine Bottles*. Peregrine Smith, Salt Lake City, Utah.
Finlay, Robert
2004 How Not to (Re)Write World History: Gavin Menzies and the Chinese Discovery of America. *Journal of World History* 15(2):229–242.
Ford, Corey
1966 *Where the Sea Breaks Its Back: The Epic Story of Early Naturalist Georg Steller and the Russian Exploration of Alaska*. Little, Brown, Boston.
Freehling, William W.
2007 *The Road to Disunion: Secessionists Triumphant, 1834–1861*. Oxford University Press, Oxford.

Frost, Frank J.

1982 The Palos Verdes Chinese Anchor Mystery. *Archaeology*, Jan.–Feb., 22–28.

Frost, Orcutt

2003 *Bering: The Russian Discovery of America*. Yale University Press, New Haven, Connecticut.

Garcia, Tracy

2007 The Smallest Island with the Oldest History: An Archaeological Investigation of CA-SMI-608 West, an Early Holocene Site on San Miguel Island, California. Unpublished senior thesis, Yale University, New Haven, Connecticut.

Gentry, Curt

1964 *The Madams of San Francisco: An Irreverent History of the City by the Golden Gate*. Ballantine Books, New York.

Gernet, Jacques

1982 *A History of Chinese Civilization*. Cambridge University Press, Cambridge.

Gibson, James R.

1992 *Otter Skins, Boston Ships, and China Goods: The Maritime Fur Trade of the Northwest Coast, 1785–1841*. McGill-Queen's University Press, Montreal.

Gibson, Rev. O.

1877 *The Chinese in America*. Hitchcock and Walden, Cincinnati, Ohio.

Glassow, Michael A.

1982 Archaeological Investigations on Eastern San Miguel Island Channel Islands National Park, California. Unpublished report submitted to National Park Service Western Regional Office, San Francisco.

Goode, George Brown (editor)

1887 *The Fisheries and Fishery Industries of the United States*. Government Printing Office, Washington, D.C.

Gragson, Ted L.

2005 Time in Service to Historical Ecology. *Ecological and Environmental Anthropology* 1(1):2–9.

Grayson, Donald K.

1990 Donner Party Deaths: A Demographic Assessment. *Journal of Anthropological Research* 46(3):223–242.

Greenwood, Roberta

1993 Old Approaches and New Directions: Implications for Future Research. In *Hidden Heritage: Historical Archaeology of the Overseas Chinese*, edited by Priscilla Wegars, pp. 375–403. Baywood, Amityville, New York.

1996 *Down by the Station: Los Angeles Chinatown, 1880–1933*. Monumenta Archaeologica No. 19. Institute of Archaeology, University of California, Los Angeles.

Gust, Sherri M.

1993 Animal Bones from Historic Urban Chinese Sites: A Comparison of Sacramento Woodland, Tucson, Ventura, and Lovelock. In *Hidden Heritage:*

Historical Archaeology of the Overseas Chinese, edited by Priscilla Wegars, pp. 177–212. Baywood, Amityville, New York.

Gyory, Andrew

1998 *Closing the Gate: Race, Politics, and the Chinese Exclusion Act.* University of North Carolina Press, Chapel Hill.

Haaker, Peter L.

2001 Abalone. In *California's Living Marine Resources: A Status Report*, edited by W.S. Leet, C.M. Dewees, R. Klingbeil, and E.J. Larson, pp. 89–97. California Department of Fish and Game, Sacramento.

Haaker, Peter L., Daniel V. Richards, Carolyn S. Friedman, Gary E. Davis, David O. Parker, and Heidi A. Togstad

1992 Mass Mortality and Withering Syndrome in Black Abalone, *Haliotis cracherodii*, in California. In *Abalone of the World: Biology, Fisheries and Culture*, edited by S.A. Shephard, M.J. Tegner, and S.A. Guzmán del Próo, pp. 214–224. Blackwell Scientific, Oxford.

Hamm, D.E., and R.S. Burton

2000 Population Genetics of Black Abalone, *Haliotis cracherodii*, along the Central California Coast. *Journal of Experimental Marine Biology and Ecology* 254:235–247.

Hansen, Richard, and Gladys Hansen

2013 *1906 San Francisco Earthquake.* Images of America. Arcadia Publishing, Charleston, South Carolina.

Hardin, Garrett

1968 The Tragedy of the Commons. *Science* 162(3859):1243–1248.

Higham, Charles

2005 East Asian Agriculture and Its Impact. In *The Human Past: World Prehistory and the Development of Human Societies*, edited by C. Scarre, pp. 234–263. Thames and Hudson, London.

Hittell, John S.

1878 *A History of the City of San Francisco and Incidentally of the State of California.* A.L. Bancroft, San Francisco.

Hobday, Alistair J., Mia J. Tegner, and Peter L. Haaker

2001 Over-exploitation of a Broadcast Spawning Marine Invertebrate: Decline of the White Abalone. *Reviews in Fish Biology and Fisheries* 10(4): 493–514.

Hoe, Susanna, and Derek Roebuck

1999 *The Taking of Hong Kong: Charles and Clara Elliot in China Waters.* Curzon Press, Richmond, Surrey, U.K.

Holliday, J.S.

1981 *The World Rushed In: The California Gold Rush Experience: An Eyewitness Account of a Nation Heading West.* Simon and Schuster, New York.

Hsu, Madeline

2000 *Dreaming of Gold, Dreaming of Home: Transnationalism and Migration between the United States and South China, 1882–1943*. Stanford University Press, Stanford, California.

Hudson, Travis, and Thomas C. Blackburn

1982 *The Material Culture of the Chumash Interaction Sphere: Volume I, Food Procurement and Transportation*. Ballena Press/Santa Barbara Museum of Natural History, Los Altos, California.

Ingstad, Helge, and Anne S. Ingstad

2000 *The Viking Discovery of America: The Excavation of a Norse Settlement in L'Anse aux Meadows, Newfoundland*. Breakwater Books, St. John's, Newfoundland.

Jackson, Jeremy B.C.

1997 Reefs since Columbus. *Coral Reefs* 16(1):S23–S32.

Jackson, Jeremy B.C., and Karen E. Alexander

2011 Introduction: The Importance of Shifting Baselines. In *Shifting Baselines: The Past and the Future of Ocean Fisheries*, edited by J.B.C. Jackson, K.E. Alexander, and E. Sala, pp. 1–7. Island Press, Washington, D.C.

Jackson, Jeremy B.C., Karen E. Alexander, and Enric Sala (editors)

2011 *Shifting Baselines: The Past and the Future of Ocean Fisheries*. Island Press, Washington, D.C.

Jackson, Jeremy B.C., Michael X. Kirby, Wolfgang H. Berger, Karen A. Bjorndal, Louis W. Botsford, Bruce J. Bourque, Roger H. Bradbury, Richard Cooke, Jon Erlandson, James A. Estes, Terence P. Hughes, Susan Kidwell, Carina B. Lange, Hunter S. Lenihan, John M. Pandolfi, Charles H. Peterson, Robert S. Steneck, Mia J. Tegner, and Robert R. Warner

2001 Historical Overfishing and the Recent Collapse of Coastal Ecosystems. *Science* 293(5530):629–637.

Jaffa, M.E.

1901 *Nutrition Investigations among Fruitarians and Chinese*. Bulletin No. 107. U.S. Department of Agriculture, Office of Experiment Stations, Washington, D.C.

Jazwa, Christopher S., Douglas J. Kennett, and Danielle Hanson

2012 Late Holocene Subsistence Change and Marine Productivity on Western Santa Rosa Island, Alta California. *California Archaeology* 4(1): 69–98.

Jeffrey, Julie Roy

1998 *Frontier Women: "Civilizing" the West, 1840–1880*. Rev. ed. Hill and Wang, New York.

Jones, Glenn A.

2008 "Quite the Choicest Protein Dish": The Costs of Consuming Seafood in American Restaurants, 1850–2006. In *Oceans Past: Management Insights from the History of Marine Animal Populations*, edited by D.J. Starkey, P. Holm, and M. Barnard, pp. 47–76. Earthscan, London.

Jones, Martin K., and Xinyi Liu

2009 Origins of Agriculture in East Asia. *Science* 324:730–731.

Jones, Mary Ellen
1998 *Daily Life on the 19th-Century American Frontier*. Greenwood Press, Westport, Connecticut.
Jones, Olive, and Catherine Sullivan
1985 *The Parks Canada Glass Glossary*. Studies in Archaeology, Architecture, and History. National Historic Parks and Sites, Canadian Parks Service Environment, Ottawa, Ontario.
Jordan, David Starr
1887 The Chinese Fishermen of the Pacific Coast. In *Notes from Fisheries and Fishery Industries of the U.S., Section II and IV*, edited by G. Goode and J. Collins. U.S. Government Printing Office, Washington, D.C.
Kalvass, Peter E., and Laura Rogers-Bennett
2001 Red Sea Urchin. In *California's Living Marine Resources: A Status Report*, edited by W.S. Leet, C.M. Dewees, R. Klingbeil, and E.J. Larson, pp. 101–104. California Department of Fish and Game, Sacramento.
Kane, Daniel
2006 *The Chinese Language: Its History and Current Usage*. Tuttle Publishing, North Clarendon, Vermont.
Karpov, Konstantin A., Peter L. Haaker, Ian K. Taniguchi, and Laura Rogers-Bennett
2000 Serial Depletion and the Collapse of the California Abalone (*Haliotis* spp.) Fishery. In *Workshop on Rebuilding Abalone Stocks in British Columbia*, edited by A. Campbell, pp. 11–24. Canadian Special Publication on Fisheries and Aquatic Sciences No. 130. Ottawa, Ontario.
Keightley, David N.
2000 *Ancestral Landscape: Time, Space, and Community in Late Shang China, ca. 1200–1045* BC. Institute of East Asian Studies, University of California, Berkeley.
Kemble, John Haskell
1953 Andrew Wilson's "Jottings" on Civil War California. *California Historical Society Quarterly* 32(3):209–224.
Kennett, Douglas J.
2005 *The Island Chumash: Behavioral Ecology of a Maritime Society*. University of California Press, Berkeley.
King, Chester D.
1990 *Evolution of Chumash Society: A Comparative Study of Artifacts Used for Social System Maintenance in the Santa Barbara Channel Region before* AD *1804*. Garland Publishing, New York.
Kislev, Mordechai E., Anat Hartmann, and Ofer Bar-Yosef
2006 Early Domesticated Fig in the Jordan Valley. *Science* 312(5778):1372–1374.
Krautkramer, R.A.
2002 Historical Archaeology of the Chinese Abalone Industry on San Nicolas Island, California. Unpublished Master's thesis, University of California, Los Angeles.

Kurlansky, Mark
1997 *Cod: A Biography of the Fish That Changed the World*. Penguin Books, New York.

Lehner, Joseph Charles
1915 *World's Fair Menu and Recipe Book: A Collection of the Most Famous Menus Exhibited at the Panama-Pacific International Exposition*. Lehner and Sefert Publishing, San Francisco.

Leighton, David, and Richard A. Boolootian
1963 Diet and Growth in the Black Abalone, *Haliotis cracherodii*. *Ecology* 44(2):228–238.

Lepofsky, Dana
2009 The Past, Present, and Future of Traditional Resource and Environmental Management. *Journal of Ethnobiology* 29(2):161–166.

Levitan, Don R., Mary A. Sewell, and Fu-Shiang Chia
1992 How Distribution and Abundance Influence Fertilization Success in the Sea Urchin *Strongylocentotus franciscanus*. *Ecology* 73(1):248–254.

Liestman, Daniel
1998 Nineteenth-Century Chinese and the Environment of the Pacific Northwest. *Pacific Northwest Quarterly* 90(a):17–29.

Lightfoot, Kent G.
2005 *Indians, Missionaries, and Merchants: The Legacy of Colonial Encounters on the California Frontiers*. University of California Press, Berkeley.

Lightfoot, Kent G., Thomas A. Wake, and Ann M. Schiff
1991 *The Archaeology and Ethnohistory of Fort Ross, California*. Archaeological Research Facility, University of California, Berkeley.

Lipsky, William
2005 *San Francisco's Panama-Pacific International Exposition*. Images of America. Arcadia Publishing, Charleston, South Carolina.

Lister, Florence C., and Robert H. Lister
1989 *The Chinese of Early Tucson: Historic Archaeology from the Tucson Urban Renewal Project*. Anthropological Papers of the University of Arizona No. 52. University of Arizona Press, Tucson.

Liu, Haiming
1996 Between China and America: The Trans-Pacific History of the Chang Family. Unpublished Ph.D. dissertation, University of California, Irvine.
2002 The Social Origins of Early Chinese Immigrants: A Revisionist Perspective. In *The Chinese in America: A History from Gold Mountain to the New Millennium*, edited by S.L. Cassel, pp. 21–36. Altamira Press, Walnut Creek, California.

Liu, Li
2004 *The Chinese Neolithic: Trajectories to Early States*. Cambridge University Press, Cambridge.

Liu, Li, Sheahan Bestel, Jinming Shi, Yanhua Song, and Xingcan Chen
2013 Paleolithic Human Exploitation of Plant Foods during the Last Glacial

Maximum in North China. *Proceedings of the National Academy of Sciences* 110(14):5380–5385.

Liu, Li, and Xingcan Chen

2012 *The Archaeology of China from the Late Paleolithic to the Bronze Age.* Cambridge University Press, Cambridge.

Liu, Li, Gyoung-Ah Lee, Liping Jiang, and Juzhong Zhang

2007 Evidence for the Early Beginning (c. 9000 cal. BP) of Rice Domestication in China: A Response. *The Holocene* 17(8):1059–1068.

Livingstone, Dewey S.

2006 *Ranches in the Sea: A History of the Islands within Channel Islands National Park.* Historic Resources Study, National Park Service, Channel Islands National Park, California.

Love, Milton S., Mary Yoklavich, and Lyman Thorsteinson

2002 *The Rockfishes of the Northeast Pacific.* University of California Press, Berkeley.

Lubbock, Basil

1927 *The Last of the Windjammers.* Brown, Son and Ferguson, Glasgow, Scotland.

Lundy, A.L. "Scrap"

1997 *The California Abalone Industry—A Pictorial History.* Best Publishing, Flagstaff, Arizona.

Lunn, Kevin

1981 Identification and Dating of Lea and Perrins' Worcestershire Sauce Bottles on Canadian Historic Sites: Interpretations Past and Present. *Canadian Journal of Archaeology* 5:1–17.

Lydon, Sandy

1985 *Chinese Gold: The Chinese in the Monterey Bay Region.* Capitola Book Co., Capitola, California.

Ma Huan

1970 *Ying-Yai Sheng-Lan: The Overall Survey of the Ocean's Shores* [1433]. Cambridge University Press, Cambridge.

MacNeish, Richard S., Geoffrey Cunnar, Zhijun Zhao, and Jane G. Libby

1998 *Re-revised Second Annual Report of the Sino-American Jiangxi (PRC) Origin of Rice Project, SAJOR.* Andover Foundation for Archaeological Research, Andover, Massachusetts.

Magnusson, Magnus, and Hermann Pálsson (translators)

1965 *The Vinland Sagas: The Norse Discovery of America.* Penguin Books, New York.

Mark, Diane Mel Lin, and Ginger Chih

1982 *A Place Called Chinese America.* Kendall/Hunt Publishing, Dubuque, Iowa.

McClenachan, Loren

2009 Documenting the Loss of Large Trophy Fish from the Florida Keys with Historical Photographs. *Conservation Biology* 23(3):636–643.

McCunn, Ruthanne Lum
1996 *Chinese American Portraits: Personal Histories, 1828–1988*. University of Washington Press, Seattle.
McEvoy, Arthur F.
1986 *The Fishermen's Problem: Ecology and Law in California Fisheries, 1850–1890*. Cambridge University Press, Cambridge.
McIntosh, Gregory C.
2000 *The Piri Reis Map of 1513*. University of Georgia Press, Athens.
McKusick, Marshall, and Erik Wahlgren
1980 The Norse Penny Mystery. *Archaeology of Eastern North America* 8:1–10.
McKusick, Marshall B., and Claude N. Warren
1959 Introduction to San Clemente Island Archaeology. *Archaeological Survey Annual Report* 1:106–183. University of California, Los Angeles.
Menzies, Gavin
2002 *1421: The Year China Discovered America*. Harper Perennial, New York.
2003 The Chinese Discovered America, Author Claims: Book Says Voyagers Preceded Columbus by 70 Years. Interview with Bob Edwards, *Morning Edition*, January 8. http://www.npr.org/templates/story/story.php?storyId=907602.
Miller, Robert J.
2006 *Native America, Discovered and Conquered: Thomas Jefferson, Lewis Clark, and Manifest Destiny*. Praeger, Westport, Connecticut.
Milner, Clyde A. II, Carol A. O'Connor, and Martha A. Sandweiss
1994 *The Oxford History of the American West*. Oxford University Press, Oxford.
Miner, C. Melissa, Jessica M. Altstatt, Peter T. Raimondi, and Todd E. Minchinton
2006 Recruitment Failure and Shifts in Community Structure Following Mass Mortality Limit Recovery Prospects of Black Abalone. *Marine Ecology Progress Series* 327:107–117.
Morris, Don P.
1994 Santa Rosa Island Survey. Paper presented at the Fourth California Islands Symposium, Santa Barbara Museum of Natural History, Santa Barbara, California.
Mullaney, Thomas S.
2010 Seeing for the State: The Role of Social Scientists in China's Ethnic Classification Project. *Asian Ethnicity* 11(3):325–342.
Muller, Fred
1987 Asian Tz'u: Porcelain for the American Market. In *Wong Ho Leun: An American Chinatown, Vol. 2, Archaeology*, edited by Great Basin Foundation, pp. 259–312. Great Basin Foundation, Riverside, California.
Munsey, Cecil
2006 The Oldest "3-IN-ONE-OIL" Container. *Bottles and Extras*, Fall, 75.

Nash, Robert Alan

1973 The Chinese Shrimp Fishery in California. Unpublished Ph.D. dissertation, University of California, Los Angeles.

Neuman, Melissa, Brian Tissot, and Glenn VanBlaricom

2010 Overall Status and Threats Assessment of Black Abalone (*Haliotis cracherodii* Leach, 1814) Populations in California. *Journal of Shellfish Research* 29:577–586.

Ogden, Adele

1941 *The California Sea Otter Trade, 1784–1884*. University of California Press, Los Angeles.

O'Meara, James

1884 The Chinese in Early Days. *Overland Monthly* 3:477–481.

Orensanz, J.M. (Lobo), Janet Armstrong, David Armstrong, and Ray Hilborn

1998 Crustacean Resources Are Vulnerable to Serial Depletion—The Multifaceted Decline of Crab and Shrimp Fisheries in the Great Gulf of Alaska. *Reviews in Fish Biology and Fisheries* 8:117–176.

Paddison, Joshua (editor)

1999 *A World Transformed: Firsthand Accounts of California before the Gold Rush*. Heyday Books, Berkeley, California.

Pan, Lynn

1994 *Sons of the Yellow Emperor: A History of the Chinese Diaspora*. Kodansha America, New York.

Pauly, Daniel

1995 Anecdotes and the Shifting Baseline Syndrome of Fisheries. *Trends in Ecology and Evolution* 10(10):430.

Pennington, J. Timothy

1985 The Ecology of Fertilization of Echinoid Eggs: The Consequences of Sperm Dilution, Adult Aggregation, and Synchronous Spawning. *Biological Bulletin* 169(2):417–430.

Perkins, Williams

1964 *Three Years in California: William Perkins' Journal of Life at Sonora, 1849–1852*. University of California Press, Berkeley.

Peterson, Curt D., Scott S. Williams, Kenneth M. Cruikshank, and John R. Dubè

2011 Geoarchaeology of the Nehalem Spit: Redistribution of Beeswax Galleon Wreck Debris by Cascadia Earthquake and Tsunami (~AD 1700), Oregon, USA. *Geoarchaeology: An International Journal* 26(2):219–244.

Pierson, Larry J., and James R. Moriarty

1980 Stone Anchors: Asiatic Shipwrecks off the California Coast. *Anthropological Journal of Canada* 18(13):17–23.

1981 New Evidence of Asiatic Shipwrecks off the California Coast. In *Underwater Archaeology: The Challenge before Us: Proceedings of the Twelfth Conference on Underwater Archaeology*, edited by G.P. Watt Jr., pp. 87–95. Special Publication No. 2. Fathom Eight, San Marino, California.

Pitcher, Tony J.
2001 Fisheries Managed to Rebuild Ecosystems? Reconstructing the Past to Salvage the Future. *Ecological Applications* 1(2):601–617.

Preston, William L.
1996 Serpent in Eden: Dispersal of Foreign Diseases into Pre-mission California. *Journal of California and Great Basin Anthropology* 18(1):2–37.
2002 Portents of Plague from California's Protohistoric Period. *Ethnohistory* 49(1):69–121.

Prince, J.D., T.J. Sellers, W.B. Ford, and S.R. Talbot
1987 Experimental Evidence for Limited Dispersal of Haliotid Larvae (Genus *Haliotis*; Mollusca: Gastropoda). *Journal of Experimental Marine Biology and Ecology* 106(3):243–263.
1988 Confirmation of a Relationship between Localized Abundance of Breeding Stock and Recruitment for *Haliotis rubra* Leach (Mollusca: Gastropoda). *Journal of Experimental Marine Biology and Ecology* 122(2):91–104.

Psota, Sunshine
1999 Clothing and Clothing Fasteners. In *Building the Los Angeles Aqueduct: Archaeological Data Recovery at the Alabama Gates Construction Camp*, edited by Thad M. Van Bueren, Judith Marvin, Sunshine Psota, and Michael Stoyka, pp. 98–119. California Department of Transportation, District 6, Fresno.

Raimondi, Peter T.
1990 Patterns, Mechanisms, Consequences of Variability in Settlement and Recruitment of an Intertidal Barnacle. *Ecological Monographs* 60(3):283–309.

Raimondi, Peter T., Michael C. Wilson, Richard F. Ambrose, John M. Engle, and Todd Erskin Minchinton
2002 Continued Declines of Black Abalone along the Coast of California: Are Mass Mortalities Related to El Niño Events? *Marine Ecology Progress Series* 242:143–152.

Ramenofsky, Ann F.
1987 *Vectors of Death: The Archaeology of European Contact.* University of New Mexico Press, Albuquerque.

Ramsey, S. Robert
1987 *The Languages of China.* Princeton University Press, Princeton, New Jersey.

Rarick, Ethan
2008 *Desperate Passage: The Donner Party's Perilous Journey West.* Oxford University Press, Oxford.

Redman, Charles L.
1999 *Human Impact on Ancient Environments.* University of Arizona Press, Tucson.

Reinman, Fred M., and Gloria A. Lauter
1984 San Nicolas Island Cultural Resources Survey Report. Unpublished

report by F.M. Reinman and Associates for Pacific Missile Test Center, Point Mugu, California.

Rice, Richard B., William A. Bullough, Richard J. Orsi, and Mary Ann Irwin
1988 *The Elusive Eden: A New History of California*. 4th ed. McGraw-Hill, New York.

Rick, Torben C., Robert L. DeLong, Jon M. Erlandson, Todd J. Braje, Terry L. Jones, Douglas J. Kennett, Thomas A. Wake, and Phillip L. Walker
2009 A Trans-Holocene Archaeological Record of Guadalupe Fur Seals (*Arctocephalus townsendi*) on the California Coast. *Marine Mammal Science* 25(2):487–502.

Rick, Torben C., and Jon M. Erlandson (editors)
2008 *Human Impacts on Ancient Marine Ecosystems: A Global Perspective*. University of California Press, Berkeley.

Rick, Torben C., Jon M. Erlandson, René Vellanoweth, and Todd J. Braje
2005 From Pleistocene Mariners to Complex Hunter-Gatherers: The Archaeology of the California Channel Islands. *Journal of World Prehistory* 19:169–228.

Rick, Torben C., and Rowan Lockwood
2013 Integrating Paleobiology, Archaeology, and History to Inform Biological Conservation. *Conservation Biology* 27(1):45–54.

Ries, Ed
1997 *Tales of the Golden Years of California Ocean Fishing, 1900–1950*. Friends of the Los Angeles Maritime Museum and the Los Angeles Maritime Museum Research Society, Los Angeles.

Roberts, Callum
2007 *The Unnatural History of the Sea*. Island Press, Washington, D.C.

Rogers-Bennett, Laura, Peter L. Haaker, Tonya O. Huff, and Paul K. Dayton
2002 Estimating Baseline Abundances of Abalone in California for Restoration. *California Cooperative Oceanic Fisheries Investigations Report* 43:97–111.

Rohrbough, Malcolm J.
1998 *Days of Gold: The California Gold Rush and the American Nation*. University of California Press, Berkeley.

Sagan, Carl
1996 *The Demon-Haunted World: Science as a Candle in the Dark*. Ballantine Books, New York.

Sala, Enric, and Jeremy B.C. Jackson
2011 Lessons from Coral Reefs. In *Shifting Baselines: The Past and Future of Ocean Fisheries*, edited by Jeremy B.C. Jackson, Karen E. Alexander, and Enric Sala, pp. 193–203. Island Press, Washington, D.C.

Sandmeyer, Elmer Clarence
1991 *The Anti-Chinese Movement in California*. University of Illinois Press, Urbana.

Saxton, Alexander
1971 *Indispensable Enemy: Labor and the Anti-Chinese Movement.* University of California Press, Berkeley.
Schultz, Peter D.
1988 Excavation of a Brickwork Feature at a Nineteenth-Century Chinese Shrimp Camp in San Francisco Bay. *Northwest Anthropological Research Notes* 22(1):121–134.
Schwartz, Steven J., and Kelly A. Rossbach
1993 A Preliminary Survey of Historic Sites on San Nicolas Island. *Proceedings of the Society for California Archaeology* 6:189–198.
Scott, Mel
1985 *The San Francisco Bay Area: A Metropolis in Perspective.* 2nd ed. University of California Press, Berkeley.
Shaw, William
1851 *Golden Dreams and Waking Realities: Being the Adventures of a Gold-Seeker in California and the Pacific Islands.* Smith, Elder and Co., London.
Shermer, Michael
1997 *Why People Believe Weird Things: Pseudoscience, Superstition, and Other Confusions of Our Time.* Henry Holt, New York.
Siu, Paul C.P.
1987 *The Chinese Laundryman: A Study of Social Isolation.* New York University Press, New York.
Smith, Bruce D.
2001 Low-Level Food Production. *Journal of Archaeological Research* 9(1):1–43.
Spence, Jonathan D.
1990 *The Search for Modern China.* W.W. Norton, New York.
1996 *God's Chinese Son: The Taiping Heavenly Kingdom of Hong Xiuquan.* W.W. Norton, New York.
Starkey, David J., Poul Holm, and Michaela Barnard (editors)
2008 *Oceans Past: Management Insights from the History of Marine Animal Populations.* Earthscan, London.
Stirling, W.G.
1913 *Opium Smoking among the Chinese.* Times of Malaya Press.
Storti, Craig
1991 *Incident at Bitter Creek: The Story of the Rock Springs Chinese Massacre.* Iowa State University Press, Ames.
Sullivan, P.J., J.M. Acheson, P.L. Angermeier, T. Faast, J. Flemma, C.M. Jones, E.E. Knudsen, T.J. Minello, D.H. Secor, R. Wunderlich, and B.A. Zanetell
2006 Defining and Implementing Best Available Science for Fisheries and Environmental Science, Policy, and Management. *Fisheries* 31(9):460–465.
Sung, Betty Lee
1967 *Mountain of Gold: The Story of the Chinese in America.* Macmillan, New York.

Sutherland, Patricia D.

2000 The Norse and Native North Americans. In *Vikings: The North Atlantic Saga*, edited by W.W. Fitzhugh and E.I. Ward, pp. 238–247. Smithsonian Institution Press, Washington, D.C.

Swetnam, Thomas W., Craig D. Allen, and Julio L. Betancourt

1999 Applied Historical Ecology: Using the Past to Manage for the Future. *Ecological Applications* 9(4):1189–1206.

Tegner, Mia J.

2000 Abalone (*Haliotis* spp.) Enhancement in California: What We've Learned and Where We Go from Here. *Canadian Special Publications of Fisheries and Aquatic Sciences* 130:61–71.

Tegner, Mia J., John D. DeMartini, and Konstantin A. Karpov

1992 The California Red Abalone Fishery: A Case Study in Complexity. In *Abalone of the World: Biology, Fisheries and Culture*, edited by S.A. Shepherd, M.J. Tegner, and S. Guzmán del Próo, pp. 370–383. Blackwell Scientific, Oxford.

Thompson, Damian

2008 *Counterknowledge: How We Surrendered to Conspiracy Theories, Quack Medicine, Bogus Science, and Fake History*. W.W. Norton, New York.

Thomson, John

1898 *Through China with a Camera*. A. Constable, Westminster, U.K.

Thornton, Russell

1987 *American Indian Holocaust and Survival: A Population History since 1492*. University of Oklahoma Press, Norman.

Tidwell, James H., and Geoff L. Allan

2001 Fish as Food: Aquaculture's Contribution; Ecological and Economic Impacts and Contributions of Fish Farming and Capture Fisheries. *EMBO Reports* 2(11):958–963.

Todd, John

1870 *The Sunset Land; or, The Great Pacific Slope*. Lee and Shepard, Boston.

Tompkins, Walker A.

1966 *Goleta: The Good Land*. Pioneer Publishing, Fresno.

Tong, Benson

1994 *Chinese Prostitutes in Nineteenth-Century San Francisco*. University of Oklahoma Press, Norman.

Toulouse, Julian Harrison

1971 *Bottle Makers and Their Marks*. Thomas Nelson, New York.

Tung, William L.

1974 *The Chinese in America, 1820–1973: A Chronology and Fact Book*. Oceana Publications, New York.

Twitchet, Denis, and Michael Loewe (editors)

1986 *The Cambridge History of China, Vol. 1: The Ch'in and Han Empires, 221 BC–AD 220*. Cambridge University Press, Cambridge.

Valentine, David

2002 Chinese Placer Mining in the United States: An Example from American Canyon, Nevada. In *The Chinese in America: A History from Gold Mountain to the New Millennium*, edited by S.L. Cassel, pp. 37–53. Altamira Press, Walnut Creek, California.

Van Tilburg, Hans K.

2007 *Chinese Junks on the Pacific: Views from a Different Deck*. University of Florida Press, Gainesville.

Vilchis, L. Ignacio, Mia J. Tegner, James D. Moore, Carolyn S. Friedman, Kristin L. Riser, Thea T. Robbins, and Paul K. Dayton

2005 Ocean Warming Effects on Growth, Reproduction, and Survivorship of Southern California Abalone. *Ecological Applications* 15(2):469–480.

Vita, Rubén, and Arnaldo Marin

2007 Environmental Impact of Capture-Based Bluefin Tuna Aquaculture on Benthic Communities in the Western Mediterranean. *Aquaculture Research* 38(4):331–339.

Voss, Barbara L.

2005 The Archaeology of Overseas Chinese Communities. *World Archaeology* 37(3):424–439.

Voss, Barbara L., and Rebecca Allen

2008 Overseas Chinese Archaeology: Historical Foundations, Current Reflections, and New Directions. *Historical Archaeology* 42(3):5–28.

Wakeman, Frederic, Jr.

1985 *The Great Enterprise: The Manchu Reconstruction of Imperial Order in Seventeenth-Century China, Volume 1*. University of California Press, Berkeley.

Walker, Phillip L., Douglas J. Kennett, Terry L. Jones, and Robert DeLong

2002 Archaeological Investigations at the Point Bennett Rookery on San Miguel Island. In *Proceedings of the Fifth California Islands Symposium*, edited by D.R. Browne, K.C. Mitchell, and H.W. Chaney, pp. 628–632. Santa Barbara Museum of Natural History, Santa Barbara, California.

Wegars, Priscilla (editor)

1993 *Hidden Heritage: Historical Archaeology of the Overseas Chinese*. Baywood Publishing, Amityville, New York.

Wheeler, Eugene D., and Robert E. Kallman

1984 *Shipwrecks, Smugglers, and Maritime Mysteries*. Pathfinder Publishing, Ventura, California.

Wiley, Walt

1975 Sir Francis Wong? Mystery of an Ancient Chinese Explorer. *Sacramento Bee*, November 30.

2001 Their Quest: A Lost Ship; Glenn County Legend Revived. *Sacramento Bee*, January 26, B1, B3.

2004 Author Calls Off Dig for Chinese Ship. *Sacramento Bee*, January 25.

Williams, Bryn
2011 *The Archaeology of Objects and Identities at the Point Alones Chinese Village, Pacific Grove, CA (1860–1906)*. Stanford University, Palo Alto, California.

Williams, Scott S.
2008 Report on 2007 Fieldwork of the Beeswax Wreck Project, Nehalem Bay, Tillamook County, Oregon. Ms. on file at Oregon State Historic Preservation Office, Salem.

Wills, John E., Jr.
1974 *Pepper, Guns, and Parleys: The Dutch East India Company and China, 1662–1681*. Cambridge University Press, Cambridge.

Wu, Cheng-Tsu (editor)
1972 *"Chink!"* World Publishing, New York.

Wylie, Jerry, and Richard E. Fike
1993 Chinese Opium Smoking Techniques and Paraphernalia. In *Hidden Heritage: Historical Archaeology of Overseas Chinese*, edited by P. Wagers, pp. 255–303. Baywood Publishing, Amityville, New York.

Yang, Jeannie K., and Virginia R. Hellmann
1998 What's in a Pot? An Emic Study of Chinese Brown Glazed Stoneware. In *Proceedings of the Society for California Archaeology* 11:59–66.

Zhao, Zhijun
2011 New Archaeobotanic Data for the Study of the Origins of Agriculture in China. *Current Anthropology* 52(S4):S295–S306.

Zhenghai, Song, and Chen Chuankang
1996 Why Did Zheng He's Sea Voyage Fail to Lead the Chinese to Make the "Great Geographic Discovery"? *Chinese Studies in the History and Philosophy of Science and Technology* 179:303–314.

Zhu, Liping
1997 *A Chinaman's Chance: The Chinese on the Rocky Mountain Mining Frontier*. University Press of Colorado, Niwot.

Zumwalt, Betty
1980 *Ketchup, Pickles, Sauces: Nineteenth Century Food in Glass*. Mark West, Fulton, California.

Index

Page numbers in *italics* refer to figures and tables.